澜湄职业教育培训中心暨柬埔寨鲁班工坊系列教材

A Series of Textbooks for Lancang-Mekong Vocational Education Training Center and Cambodia Luban Workshop

4G 通信网络管理员(中级)

4G Communication Network Administrator—Intermediate

主 编 韩 健

Chief editor：HAN Jian

副主编 刘赟宇 高 源 孔 雷 张 磊

Deputy editors：LIU Yunyu　GAO Yuan　KONG Lei　ZHANG Lei

西安电子科技大学出版社

Brief Introduction of the Content

Based on the 4G TD-LTE (Time Division-Long Term Evolution) technology, this book uses the modularity-learning mode and the TD-LTE eNodeB equipment of Datang Mobile Company to give a comprehensive instruction about the hardware structure, device opening and setting, equipment maintenance, etc. through different tasks arrangement. This book has three training parts. The first part is the TD-LTE wireless technology, the second one is the recognition of TD-LTE products, and the third one is the operating and maintenance of the base stations. Through the learning of the three training parts, the readers can have a good knowledge of the basic setting and maintenance of a 4G base station, as well as a sound understanding of the structure installation and testing of the hardware, and the starting configuration and testing of the base devices. They can understand the basic knowledge involved in running and testing the mobile communication system. The book aims to train the high-qualified talents for constructing and maintaining the 4G base station, and enables them to well grasp the running and maintaining skills for the communication base station and meet the work requirements of the standardized maintenance.

The knowledge of this book is comprehensive, and it gives a comprehensive introduction to the basic concepts of the mobile communication theory. The book has fully talked about the constructing and maintaining for a 4G base station, and used the Datang TD-LTE experimental simulation teaching and learning software to illustrate the working processes and the related tasks. Therefore, the knowledge in this book is more practical.

This book can be used as a professional material for some majors like electronics, communication or the related ones in vocational schools and colleges, and also it can serve as a reference and guidance book for professionals or technicians.

图书在版编目(CIP)数据

4G 通信网络管理员(中级) / 韩健主编. — 西安：西安电子科技大学出版社，2021.12
ISBN 978-7-5606-6043-1

Ⅰ. ①4… Ⅱ. ①韩… Ⅲ. ①计算机网络管理—高等职业教育—教材—英文
Ⅵ. ①TP393.07

中国版本图书馆 CIP 数据核字(2021)第 069923 号

策划编辑　刘玉芳
责任编辑　郑一锋　刘玉芳
出版发行　西安电子科技大学出版社(西安市太白南路 2 号)
电　　话　(029)88202421　88201467　　邮　编　　710071
网　　址　www.xduph.com　　　　电子邮箱　xdupfxb001@163.com
经　　销　新华书店
印刷单位　陕西天意印务有限责任公司
版　　次　2021 年 12 月第 1 版　　2021 年 12 月第 1 次印刷
开　　本　787 毫米×1092 毫米　　1/16　　印　张　8
字　　数　179 千字
印　　数　1～1000 册
定　　价　28.00 元
ISBN 978−7−5606−6043−1 / TP
XDUP 6345001-1
***** 如有印装问题可调换 *****

General Preface

Serving the Belt and Road Initiative of China, the Lancang-Mekong Vocational Education Training Center and Cambodia Luban Workshop is a joint project undertaken by Tianjin Sino-German University of Applied Sciences(TSGUAS) for the Ministry of Foreign Affairs, the Ministry of Education and the Tianjin Municipal People's Government. Based in Cambodia, the project is designed to serve five countries in the Lancang-Mekong area and radiate to other ten ASEAN countries. It integrates functions of vocational training, vocational education, scientific research, cultural inheritance and innovation&entrepreneurship, develops both academic and non-academic education, and operates as a market-oriented international vocational training center.

At the initial stage of the project, 18 training rooms including mechanical processing technology, electrical technology and communication technology were built in three training centers for mechatronics and communication technology majors, with a total construction area of 6,814 m^2 and more than 1,600 sets of equipment.

The project will implement a "three-phase" plan. Based on the specialty construction in the first phase, international tourism, logistics engineering, automobile maintenance, building electricity and other specialties will be set up in the second phase to carry out technical skills training for Chinese&Cambodian enterprises and Cambodian people. Meanwhile, higher vocational education, applied technology undergraduate education, joint postgraduate education and other academic educations will be carried out to explore systematic talents cultivation of "medium and high vocational education, undergraduate education, and postgraduate education for a master's and doctoral degree".

Since 2017, as many as 95 articles about the project have been published by mainstream media including People's Daily, Guangming Daily, China Education News, Xinhuanet, etc. from home and abroad. After over two months of field study and research, Tianjin Television produced two feature stories named "Khmer Training", each lasting 30 minutes. The two episodes were broadcast on May 6th and May 13th 2019 respectively, featuring "on and on sails the vocational education, overseas shines the Luban Workshop". They give a full coverage of how TSGUAS teachers brought advanced skills to local areas and how friendship flourished along the Belt and Road Initiative route—a great contribution to the BRI. On July 18th, 2019, the Royal Government of Cambodia conferred the Officer of the SAHAMETREI Medal to the Secretary of the Party Committee of TSGUAS, and the Knight of the SAHAMETREI Medal to the President and Vice President in charge of this project, with the signature of Prime Minister Hun Sen of Cambodia. On July 22, 2019, China Education Association for International Exchange awarded TSGUAS the medal of "Featured Cooperation Project of China-ASEAN Higher Vocational Colleges". In October 2019, the President of National Polytechnic Institute of

Cambodia (NPIC) presented 11 teachers with certificates and medals for their outstanding contributions to the Ministry of Labor and Vocational Training of Cambodia. Tianjin Sino-German University of Applied Sciences together with National Polytechnic Institute of Cambodia (NPIC) and their partners with enterprises was approved as the Belt and Road Joint Laboratory (Research Center)—Tianjin Sino-German and Cambodia Intelligent Motion Device and Communication Technology Promotion Center in December, 2020.

The Center has become a training base in Lancang-Mekong areas for technical talents training, a talent support base for Chinese enterprises overseas, a demonstration base for international students, and a base for teachers training. The Center is a key educational project of the Ministry of Foreign Affairs to serve the Belt and Road Initiative with foreign participation and entity institutions involved locally. The project will serve the social-economic and cultural development of the countries along the Initiative, enhancing the well-being of mankind; it will also serve the production output capacity of Chinese enterprises to help national development as well as enhance the international development of vocational education and the quality of its connotation. The project is a bridge connecting vocational education of Tianjin with the world, which marks a new stage of the city's international exchange and cooperation from a lower-medium to a medium-higher level.

The team of the project has compiled a series of textbooks for training, involving six occupations (electrotechnics, lathe, milling, CNC operation, bench and 4G communication network) from elementary, intermediate to advanced level based on current human resources situation in Lancang-Mekong countries, China's teaching equipment, and Chinese vocational qualification standards. These 19 textbooks target competence development and orient students to work tasks, combining theory with practice, and learning with practicing so as to put knowledge and skills into real situations. The textbooks aim to provide skills standards for the six occupations and lay foundations for the upgrading of the technological level of Lancang-Mekong countries.

ZHANG Xinghui

Party Secretary of Tianjin Sino-German University of Applied Sciences

March, 2021

A Series of Textbooks for Lancang-Mekong Vocational Education Training Center and Cambodia Luban Workshop Editorial Committee

Preface

The book focuses on the practice and operation, and aims to explain the related work tasks for learners and readers. The contents are based on the actual needs of the work. By using the Datang fourth generation mobile communication system for network installation, maintenance, testing and adjusting, the book has combined the training of practice skills and professional skills together, and the book organization is much newer and the content is more practical.

The book aims to train the practical talents for mobile communication technology. Based on the cooperation between universities and companies, the book has been written by the first-class teachers who have more teaching experiences and the company's senior engineers who have more engineering and practical experiences. The book can be divided into three parts according to the local learners' knowledge, and their enterprise jobs and their levels specifically as follows:

4G communication network manager-primary level (5th);

4G communication network manager-intermediate (4th);

4G communication network manager-senior level (3rd).

Primary level: it is for new workers' training, and the learners can understand the basic installation skill of the job. For example, they can work as installation engineers or installation supervisors.

Intermediate: learners can enter this level only after they finish learning the primary level and have two years' working experience. At this level, the learners will learn how to start, adjust and maintain a base station, and the corresponding jobs are the adjusting engineer, and operation and maintenance engineer.

Senior level: learners can enter this level only after they finish learning the middle level and have one years' or two years' working experience. At this level, the learners will learn the base station testing, and network planning and optimization. The corresponding jobs are the planning engineer and adjusting engineer.

The intermediate level is based on the basic LTE theory, and its practice and operation. It mainly introduces the LTE network structure, practical principles, key technologies, theoretical basis of important LTE concepts involved, as well as the signaling procedures and their related parameters and information.

As the intermediate level book, it uses the Datang Company's main product-EMB 5116 base station. According to the theory and the actual network building experiences by Chinese operation providers, the book talks comprehensively about the full network processes like LTE wireless connection, key network planning, installation and adjusting, etc. It can be a theory-practice platform for the primary learners who have some basic communication knowledge or the people who are interested in this field. We strongly recommend that the learners using the Datang simulation software V-Lab during the learning; it can not only help the reader have better understanding about the whole process for the LTE network building, and also help them gain the ability for base station operation and maintenance.

It is available to scan the two-dimensional code below for corresponding contents of this book in Chinese.

Due to the limited knowledge of the writers, helpful advices from experts and readers are warmly welcome.

Compilers

November, 2019.

译文

Contents

Training Module 1: TD-LTE Wireless Technology(16 Class Hours)

【Basic Introduction】

This part mainly focuses on introduction to the physical technologies and processing, including the technologies of multi-address, duplex mode, frame structure and physical resource, physical channel and signal, multi-antenna, and link adaptation and channel scheduling.

【Training Purposes】

1. Knowledge Purpose

(1) Understand the knowledge about multi-address technology, duplex mode, frame structure and physical resource, physical channel and signal, etc.

(2) Understand the technologies of multi-antenna, link adaptation, and channel scheduling.

2. Skill Purpose

(1) Analyze the relation between wireless technology and network.

(2) Use wireless skill to solve the network failure and finish the network optimization.

【Training Requirements】

1. Preparation for the tools, instruments and other materials

A set of mobile communication equipment.

2. Knowledge to be assessed

(1) LTE multi-address technology.

(2) LTE frame structure and physical resource.

(3) Area searching.

3. Skill evaluation

(1) Draw the impulse form and frequency spectrum of an OFDM subcarrier.

(2) Understand the link adaptation and channel scheduling technology.

(3) Draw the random connection process based on competition.

Work Task 1: LTE (Long Term Evolution) Port

1. Wireless Interface Protocol

The wireless interface is between the related terminals and the eNodeB (Evolved Node B), and it is totally open. The wireless interface protocol contains three layers, and two planes. The three layers are the physical layer, data link layer and network layer respectively; while the two planes refer to the control plane and user plane, as shown in Figure 1-1-1.

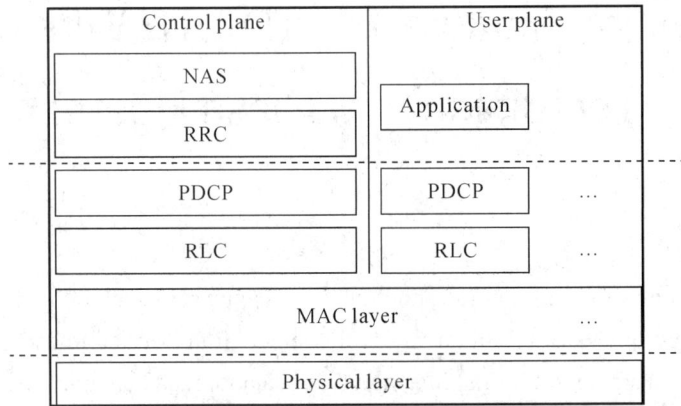

Figure 1-1-1   LTE wireless interface protocol

Layer 1: physical layer provides the wireless resource and physical processing for high-level data.

Layer 2: date link layer, including MAC(Media Access Control), RLC(Radio Link Control), and PDCP (Packet Data Convergence Protocol), which makes distinct identification and provides related service for the three layers' data.

Layer 3: network layer, including RRC(Radio Resource Control), which is the user of the wireless interface service and refers to the service data of RRC and users.

The wireless interface control surface mainly manages and controls the wireless interface, which includes RRC protocol, data link control protocol and packet data convergence protocol. NAS (Non-Access Stratum), located in the UE (User Equipment) and MME (Mobility Management Entity), controls the protocol entities and manages the non-connection parts, and eNodeB (eNB) does nothing to NAS. RRC protocol entity, located in the UE and eNodeB, is responsible for the control and management of different connection parts. Both the data link layer and the physical layer have the transmission function for RRC protocol message. All the information above is shown in Figure 1-1-2.

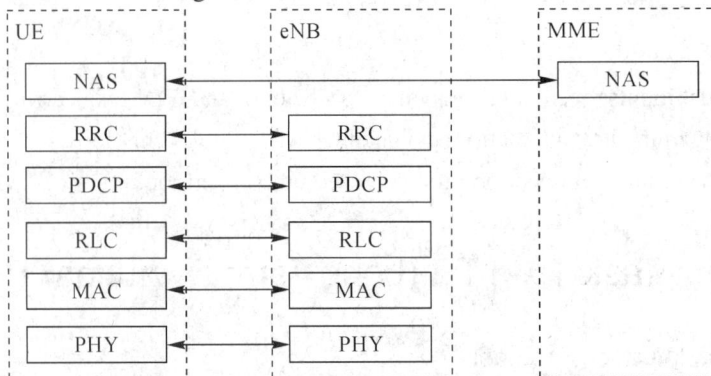

Figure 1-1-2   Wireless interface control surface

The wireless interface user platform protocol refers to the protocols of data link layer (MAC, RLC, PDCP) and physical layer. The physical layer can provide data transmission for the data link layer. The physical layer can provide the related service for the MAC sub-layer through transmission channel, and the MAC sub-layer can provide related service to the RLC sub-layer through the logic channel, as shown in Figure 1-1-3.

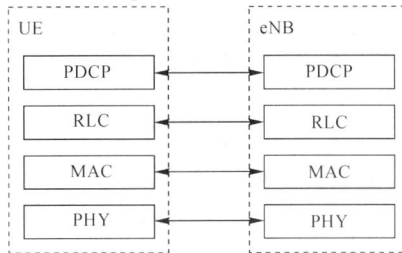

Figure 1-1-3　LTE wireless interface user platform

2. Signaling and Data streams

Figure 1-1-4 is the flowchart for the LTE data stream and the signaling stream. It can be easily recognized that different data streams work in different ways. The solid arrow shows how the data flow from mobile phone to the key network, and the dotted arrow describes how the signal from mobile phone goes to the key network.

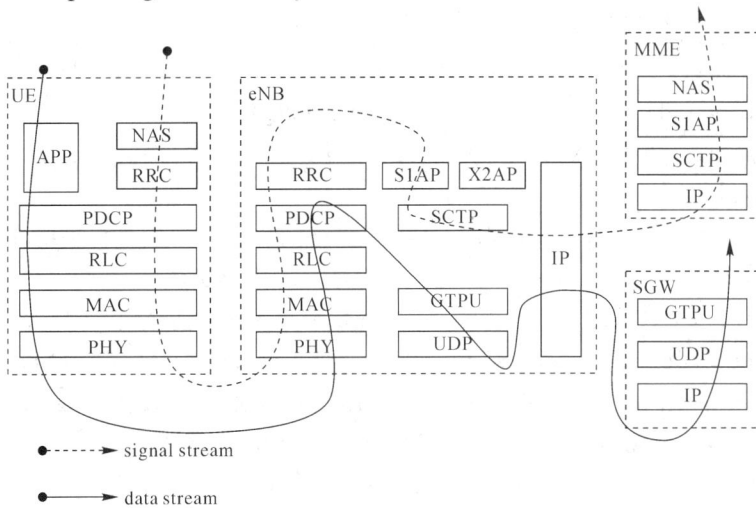

Figure 1-1-4　The operation flowchart of LTE signaling stream and data stream

3. S1 Interface Protocol

The S1 interface control surface has the control function for eNodeB and MME. It contains the application protocol and the signal load used for transmitting application protocol. The user surface of the S1 interface provides data transmission between the eNodeB and the S-GW (Serving Gate Way). The user surface contains the data load used for data stream, and the data stream here refers to the tunnel protocol of the transmission network layer.

The control surface of S1 (S1-MME) is the interface between the eNodeB and the MME. The transmission network layer is based on the IP which can transmit the signal PDU (Packet

Data Unit) from point to point. The SCTP (Stream Control Transmission Protocol) is used above the IP level, which can provide reliable transmission for the application layer. The S1AP (S1 Application Protocol) refers to the S1 Application Protocol, as shown in Figure 1-1-5.

The S1 user surface (S1-U) can transmit the user data between the eNodeB and the S-GW. The transmission layer is based on IP transmission, GTP-U (GPRS Tunneling Protocol for the user plane) is above the UDP/IP and can transmit the user PDU between the eNodeB and the S-GW, as shown in Figure 1-1-6.

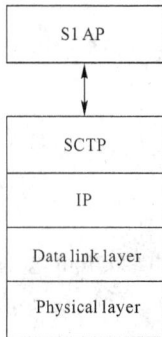

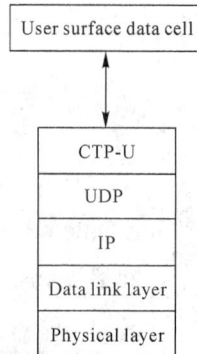

Figure 1-1-5   S1 control surface protocol (eNB-MME)   Figure 1-1-6   S1 user surface protocol

4. X2 Interface Protocol

The X2 interface is used for the connection of the eNodeBs. The definition of the X2 interface adopts the same principle that is used for defining the S1 interface. The control surface protocol and the user surface protocol of the X2 interface have the structures that are similar to those of the S1 interface.

The logic connection between the two eNodeBs decides the X2 control (X2-C). The transmission network layer is built above the SCTP of IP, and the signalling protocol of the application layer is X2AP (X2 Application Protocol), as shown in Figure 1-1-7.

The X2 user surface interface (X2-U) is defined by what between the two eNodeBs, and X2-U can provide the nonguaranteed transmission of the user PDU. The transmission network layer is based on IP, and the GTP-U protocol is above the UDP/IP and can transmit the user surface PDU between the eNodeBs. The protocol stack of the X2-U is the same as that of the S1-U, as shown in 1-1-8.

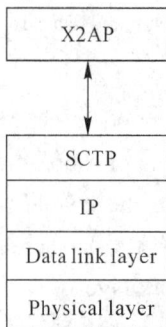

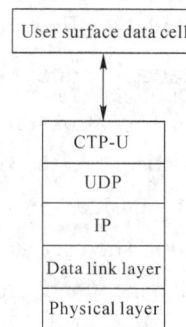

Figure 1-1-7   X2 control protocol   Figure 1-1-8   X2 user protocol

Work Task 2:   LTE Multi-address Technology

The OFDM technology of CP (Cyclic Prefix) is used for the downlink of LTE system, and the SC-FDMA (Single-carrier Frequency-Division Multiple Access) technology of CP is used for the uplink.

1. OFDM

1) OFDM (Orthogonal Frequency Division Multiplexing)

OFDM is a kind of multicarrier transmission technology. Compared with the traditional multicarrier transmission such as FDM, the OFDM can use more and narrower orthogonal subcarriers for transmission, as shown in Figure 1-2-1.

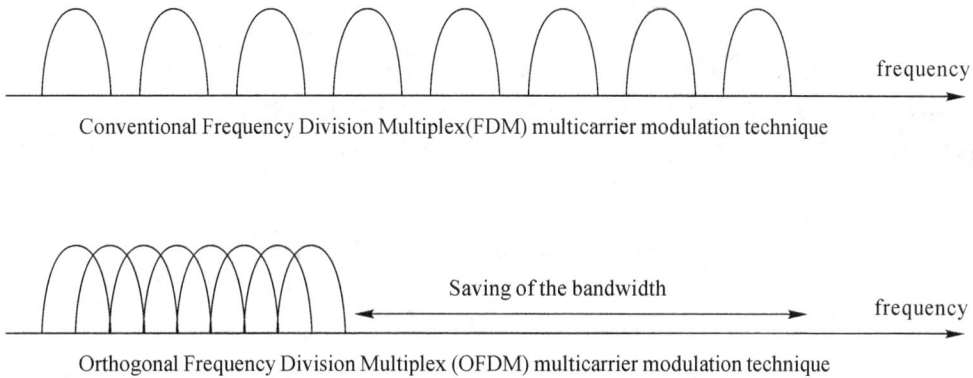

Conventional Frequency Division Multiplex(FDM) multicarrier modulation technique

Orthogonal Frequency Division Multiplex (OFDM) multicarrier modulation technique

Figure 1-2-1   Comparison of OFDM and FDM

Figure 1-2-2 has shown the pulse shape and the spectrum of an OFDM subcarrier (CP is not included).

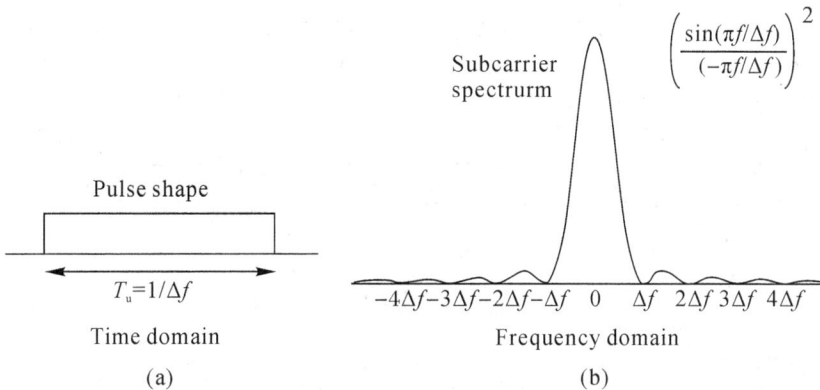

Subcarrier spectrrum

$$\left(\frac{\sin(\pi f/\Delta f)}{(-\pi f/\Delta f)}\right)^2$$

Pulse shape

$T_u = 1/\Delta f$

$-4\Delta f\ -3\Delta f\ -2\Delta f\ -\Delta f\quad 0\quad \Delta f\ \ 2\Delta f\ 3\Delta f\ 4\Delta f$

Time domain

Frequency domain

(a)

(b)

Figure 1-2-2   Pulse shape and spectrum of an OFDM subcarrier

Therefore, an OFDM consists of several subcarriers, and they have overlapped with each other and are orthogonal. The Δf means the subcarrier space, as shown in Figure 1-2-3.

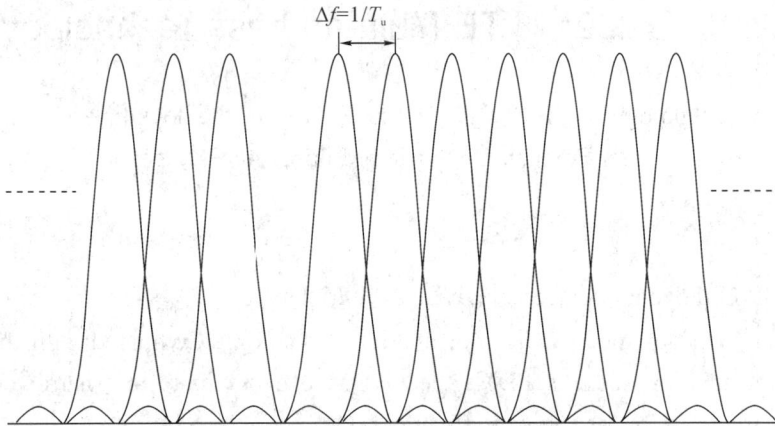

Figure 1-2-3   An OFDM freguency Spectrum

The orthogonality of OFDM means any two subcarriers in an OFDM are orthogonal to each other, and the OFDM demodulation becomes much easier because of the orthogonality. It is composed by a series of correlators, and each correlator has one subcarrier, as shown in Figure 1-2-4.

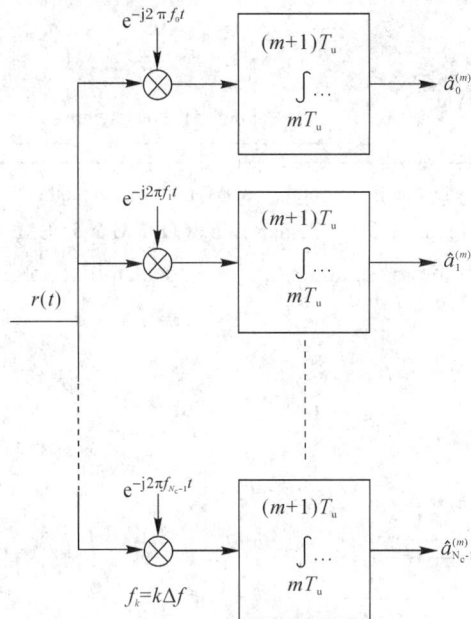

Figure 1-2-4   OFDM demodulation

2) Cycle prefix

In the OFDM system, the subcarrier orthogonality means the integer period of the complex index in one integration interval. Under the time dispersion signal channel, the orthogonality of subcarrier will be influenced more or less if the CP is not inserted in the OFDM. This is because the space for one demodulation has overlapped with the others (it can cause ISI). This means the integration interval not only contains the integer period of complex index for the main path, but also contains the fractional period of complex index for other paths, which can influence the subcarrier orthogonality. Therefore, the subcarrier has the ISI (Inter Symbol Interference), and also the ICI (Inter Code Interference) under the time dispersion signal channel, as shown in Figure 1-2-5.

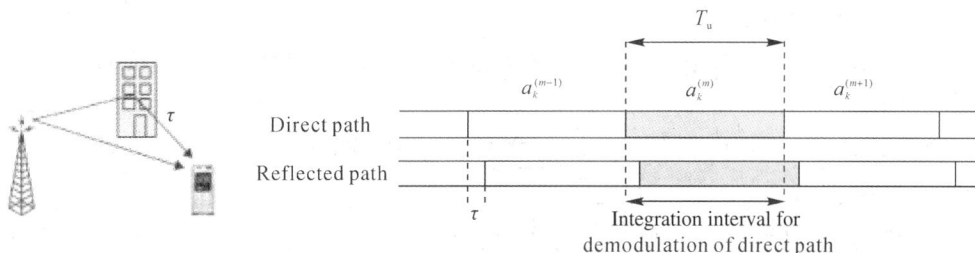

Figure 1-2-5　Formation of ISI and ICI

The use of CP can greatly reduce the interference of the related symbols and subcarriers. The insertion of CP means that the last part of CP is copied and inserted at the beginning of OFDM. Therefore, the OFDM length changes from T_u to $T_u + T_{CP}$ due to the CP insertion, and T_{CP} means the length of CP. As shown in the lower part of Figure 1-2-6, the subcarrier orthogonality can be guaranteed if the integral cycle of the receiving terminal is $T_u = 1/\Delta f$ and the CP length is larger than the time-delay extension values.

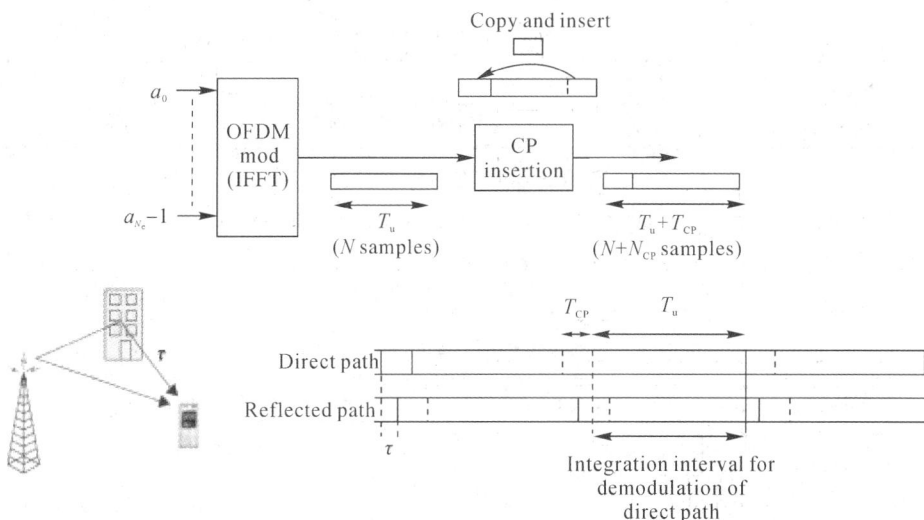

Figure 1-2-6　CP insertion

In fact, the OFDM modulation can be realized by IFFT (Inverse Fast Fourier Transform). Under this situation, the CP insertion means the last part N_{CP} of IFFT should be copied and inserted at the front, and the length of IFFT output changes from N to $N + N_{CP}$.

CP insertion can effectively reduce the OFDM sensibility to the time dispersion signal channel, and also it has some disadvantages:

(1) Power lose: only the receiving signal power of $T_u/(T_u + T_{CP})$ is really used to OFDM demodulation.

(2) Bandwidth lose: the insertion of CP can reduce the OFDM speed rate, but the

bandwidth of signal transmission has no changes.

　　3) OFDM parameters

The following OFDM parameters are for OFDM transmission:

　　(1) subcarrier space Δf;

　　(2) subcarrier number N_c and space Δf can (decide the transmission bandwidth of an OFDM signal);

　　(3) CP length T_{CP} and subcarrier space $\Delta f = 1/T_u$ (decide the time length $T = T_{CP} + T_u$ of the whole OFDM).

For the downlink transmission of LTE, the basic OFDM parameters includes the carrier space, subcarrier number and the length of cycle prefix.

The downlink of a LTE system supports two kinds of subcarrier spaces, and they are as follows:

　　① $\Delta f = 15$ kHz, is used for the transmission of unicast and MBSFN;

　　② $\Delta f = 7.5$ kHz, can only be used for the transmission of independent carrier MBSFN (Multicast Broadcast Single Frequency Network).

Subcarrier number N_c: different system bandwidth has different number of subcarriers; the LTE system has its own requirement for the number of subcarrier, as shown in Table 1-2-1.

Table 1-2-1　LTE OFDM subcarrier number N_c

Signal channel bandwidth (MHz)	1.4	3	5	10	15	20
Subcarrier number N_c	72	180	300	600	900	1200

CP length T_{CP}: For $\Delta f = 15$ kHz, the LTE supports two kinds of CP lengths: normal CP and extend CP, which are respectively used under different transmission situations; for $\Delta f = 7.5$ kHz, the LTE only supports extent CP. The lengths for the Different OFDM CPs in a time slot are different in ensuring one time slot as exactly 0.5 ms.

Table 1-2-2　LTE OFDM CP length T_{CP}

Setting	Data Name	CP length $N_{CP, l}$
Normal CP	$\Delta f = 15$ kHz	160 for ($l = 0$) 144 for ($l = 1,2,\cdots,6$)
Extent CP	$\Delta f = 15$ kHz	512 for ($l = 0,1,\cdots,5$)
	$\Delta f = 7.5$ kHz	1024 for ($l = 0,1,2$)

$N_{CP, l}$ means the CP sample value of the l OFDM in one time slot(T_{CP}).

2. SC-FDMA

LTE uplink transmission uses SC-FDMA, which is DFTS-OFDM.

　　1) DFTS-OFDM (Discrete Fourier Transform Spread OFDM) principle

The size-M modulation will first transform DFT of M points, and then size-N modulation will transform IDFT of N points; $N > M$, and the value of other points are set to 0. You need to pay attention to the fact that the value of IDFT should be $N = 2^n$, and it can realize IFFT which has the base value of 2. Similar to OFDM, CP is inserted to each transmission block, and it can easily use the lower complicated frequency domain equalization if CP is used. The transmission principle of DFTS-OFDM is shown in Figure 1-2-7.

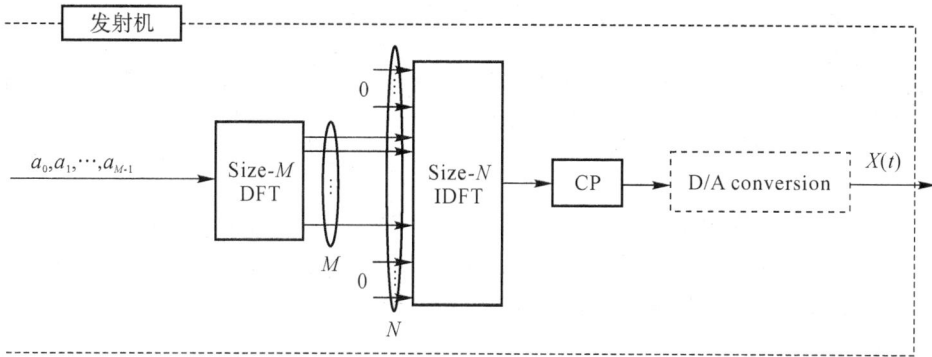

Figure 1-2-7   Formation of DFTS-OFDM signal

If the value M of DFT is the same as the value N of IDFT, the cascaded DFT and IDFT will offset each other. If $N>M$, the output signal of IDFT will be one with the characteristics of a single carrier; the signal has a low peak ratio and the signal bandwidth will be decided by M. Exactly speaking, if the output sampling rate of IDFT is f_s, the bandwidth of the transmission signal will be BW = M/N f_s theoretically. Therefore, the instant bandwidth of transmission signal can be changed by adjusting the M block value.

In order to guarantee the high flexibility of the instant bandwidth, the value M of DFT can not be promised to be 2^m, and m should be an integer. However, if M is a relatively small prime number product, DFT can still be realized by using the FFT whose low complex base number is not 2. For example, if the DFT value is $M=144$, it can be realized by using base-number-2 and base-number-3 FFT ($144 = 3^2 \times 2^4$).

There are two methods for DFT output to reflect IDFT input, and they are Localized DFTS-OFDM and Distributed DFTS-OFDM. The Localized DFTS-OFDM means that DFT output can be reflect on several continuous inputs of IDFT, and Distributed DFT-OFDM refers to that the DFT output is reflected on the several IDFT inputs which have the same space among them, and other inputs will be 0, as shown in Figure 1-2-8.

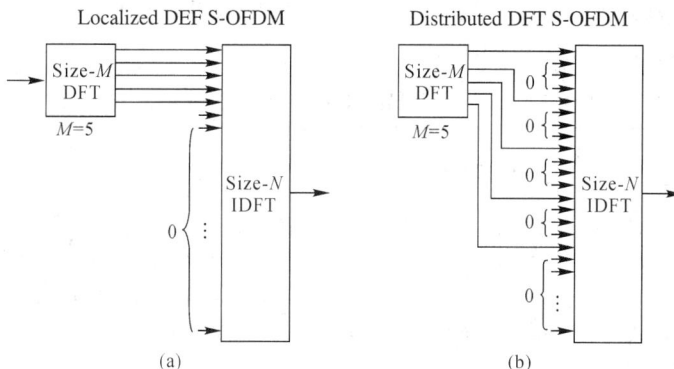

Figure 1-2-8   Localized DFT-S-OFDM and Distributed DFT-S-OFDM

Figure 1-2-9 has shown the frequency spectrums of Localized DFTS-OFDM and Distributed DFTS-OFDM signals. Although the signal frequency spectrum of Distributed DFTS-OFDM is distributed in the whole system bandwidth, and it still owns the characteristics of a single carrier. In fact, Distributed DFTS-OFDM is IFDMA.

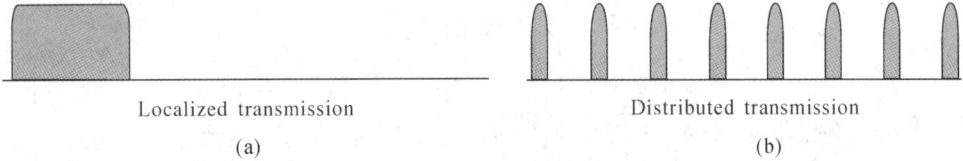

Localized transmission   Distributed transmission

(a)   (b)

Figure 1-2-9   Localized and Distributed DFTS-OFDM frequency spectrums transmission

Except the uplink scanning reference signal (SRS), you need to pay attention to the fact that LTE doesn't support to transmit signal by using Distributed DFTS-OFDM.

2) DFTS-OFDM parameters

Similar to OFDM, DFTS-OFDM also has the following parameters:

(1) subcarrier space Δf;

(2) subcarrier number N_c (decide the transmission bandwidth of DFTS-OFDM together with subcarrier space Δf);

(3) CP length T_{CP}(decide the whole DFTS-OFDM time length $T = T_{CP} + T_u$ together with subcarrier space $\Delta f = 1/T_u$).

For LTE uplink transmission, the basic DFT-OFDM parameters are as follows:

(1) Subcarrier space: LTE system uplink only supports one kind of subcarrier space, which is $\Delta f = 15$ kHz.

(2) Subcarrier number N_c: the different system bandwidth has different subcarrier numbers, and the number of subcarriers for a LTE system is shown in Table 1-2-3.

Table 1-2-3   LTE DFTS-OFDM subcarrier number N_c

Signal channel bandwidth(MHz)	1.4	3	5	10	15	20
Subcarrier number N_c	72	180	300	600	900	1200

(3) CP length T_{CP}: LTE supports two kinds of CP, and they are normal CP and extent CP. In order to guarantee one time slot is 0.5 ms, the CP length for different DFTS-OFDM in one time slot is different, as shown in Table 1-2-4.

Table 1-2-4   LTE DFTS-OFDM CP Length T_{CP}

Parameter	Data Name	CP length $N_{CP,l}$
Normal CP	$\Delta f = 15$ kHz	160 ($l = 0$) 144 ($l = 1,2,\cdots,6$)
Extent CP	$\Delta f = 15$ kHz	512 ($l = 0,1,\cdots,5$)

$N_{CP,l}$ means the CP sample value of the l DFTS-OFDM in a time slot.

Work Task 3: LTE Frame Structure and Physical Resource

1. Frame structure

LTE supports two kinds of frame structures, which are useful for the operation of FDD (Frequency Division Duplexing), H-FDD and TDD (Time Division Duplexing). The definition time unit of LTE is $T_s = 1/(15\,000 \times 2048)$ second.

1) Frame struture type 1

Frame structure type 1 can be used to the full-duplex mode or half-duplex mode. Each radio frame length is 10 ms which is made up of 20 slots, and each slot length is 0.5 ms. The time slot caption goes like $0,1,\cdots,19$. One subframe has two time slots which are next to each other, and the i-th subframe is composed of the $2i$-th time slot and the $(2i + 1)$-th time slot, as shown in Figure 1-3-1.

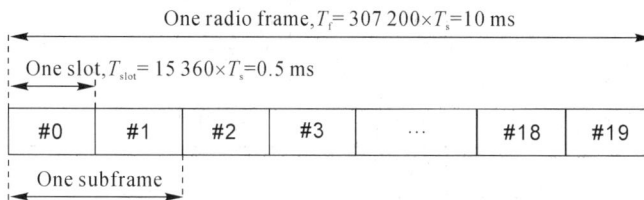

Figure 1-3-1   Frame structure type 1

For FDD, 10 subframes can be used for the downlink transmission, and also other 10 subframes for the uplink transmission, in each 10 ms. The uplink and downlink transmission can be separated on the frequency domain.

2) Frame structure type 2

Frame structure type 2 can be applied to TDD. Each frame has two half-frame, and each half-frame is 5 ms. Each half-frame has 8 slots, and each slot is 0.5 ms; and also it contains three special time slots, which are DwPTS (Downlink Pilot TimeSlot), GP (Guard Period) and UpPTS(Uplink Pilot TimeSlot). The DwPTS and UpPTS length can be set, and the total length of DwPTS, GP and UpPTS should be 1ms. Subframes 1 and 6 have DwPTS, GP and UpPTS, and other subframes have two time slots which are next to each other. The i-th subframe is composed of the $2i$-th time slot and the $(2i + 1)$-th time slot, as shown in Figure 1-3-2.

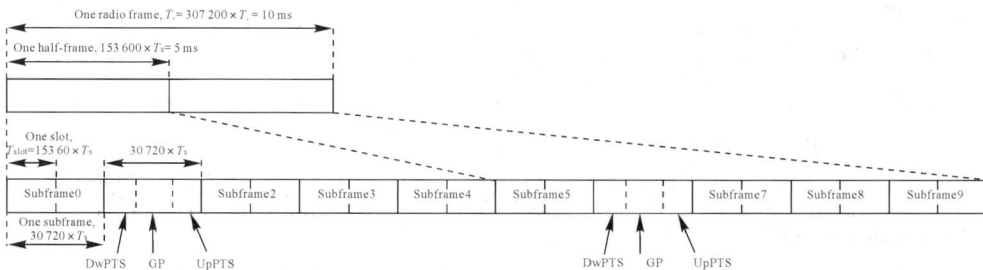

Figure 1-3-2   Frame structure type 2

Under the frame structure type 2 environment, the 5 ms and 10 ms uplink and downlink subframes can switch their period, and the setting is shown in Table 1-3-1.

Table 1-3-1   Setting of frame structure type 2

Configuration	Switch-point periodicity	Subframe number									
		0	1	2	3	4	5	6	7	8	9
0	5 ms	D	S	U	U	U	D	S	U	U	U
1	5 ms	D	S	U	U	D	D	S	U	U	D
2	5 ms	D	S	U	D	D	D	S	U	D	D
3	10 ms	D	S	U	U	U	D	D	D	D	D
4	10 ms	D	S	U	U	D	D	D	D	D	D
5	10 ms	D	S	U	D	D	D	D	D	D	D
6	10 ms	D	S	U	U	U	D	S	U	U	D

3 special time-slot settings for special subframes are shown in Table 1-3-2.

Table 1-3-2   Time-slot setting of the special subframes

Setting choice	Normal CP			Extent CP		
	DwPTS	GP	UpPTS	DwPTS	GP	UpPTS
0	3	10	1	3	8	1
1	9	4	1	8	3	1
2	10	3	1	9	2	1
3	11	2	1	10	1	1
4	12	1	1	3	7	2
5	3	9	2	8	2	2
6	9	3	2	9	1	2
7	10	2	2	—	—	—
8	11	1	2	—	—	—

2. Physical resource

The smallest resource unit used in uplink and downlink transmission is defined as the resource element (RE). The resources blocks (RB) is based on RE, and each RB has several REs.

1) Resource element

The resource unit refers to each unit of the subcarrier in each OFDM or SC-FDMA, as shown in Figure 1-3-3.

The number of the symbols in each time slot is related to the CP used. Normal CP has 7 symbols in one time slot, and extent CP has 6 symbols in one time slot, as shown in Table 1-3-3.

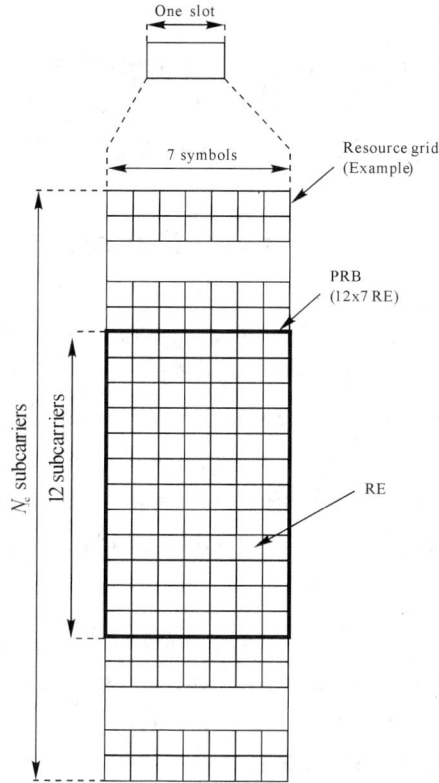

Figure 1-3-3   RE, PRB and resource grid diagram

Table 1-3-3   Symbol number in different CP time slots

Subcarrier space	CP	OFDM / SC-FDMA symbol number
$\Delta f = 15$kHz	Normal CP	7
	Extent CP	6
$\Delta f = 7.5$kHz	Extent CP	3

2) Physical resource block

If the width of the adjacent physical resource on the frequency domain is 180 kHz in one time slot, it is defined as a physical resource block (PRB), as shown in Figure 1-3-3.

If the subcarrier number and the symbol number are used to describe the relationship between PRB and RE, as shown in Table 1-1-4.

Table 1-3-4   PRB size

Subcarrier space	CP	Subcarrier number	OFDM / SC-FDMA symbol number	RE number
$\Delta f = 15$ kHz	Normal CP	12	7	84
	Extent CP	12	6	84
$\Delta f = 7.5$ kHz	Extent CP	24	3	72

3) Resource grid

All the resource units used for the signal transmission in one time slot are defined as a resource grid, which has an integral number of PRB and also can be represented by the subcarrier number, OFDM or SC-FDMA symbol number.

Work Task 4: Physical Channel and Signal

The physical channel is the real load for the senior signal in wireless environment. The subcarrier, time slot and antenna can jointly decide the physical channel in LTE, and there is a series of wireless time frequency resource (Resource Element, RE) on the special antenna. A physical channel has its own starting time, ending time and duration time. The physical channel can be continuous or non-continuous on the time domain. The continuous physical channel lasts from the starting time to the ending time. It should be clearly indicated which time slices consist of the non-continuous physical channel.

1. Downlink physical channel

LTE has six physical channels, and they are as follows:

(1) physical Downlink Shared Channel, PDSCH;

(2) physical Multicast Channel, PMCH;

(3) physical Downlink Control Channel, PDCCH;

(4) physical Broadcast Channel, PBCH;

(5) physical Control Format Indicator Channel, PCFICH;

(6) physical Hybrid ARQ Indicator Channel, PHICH.

The channel reflection refers to the relation among logic channel, transmission channel and physical channel, and this relation contains not only the service and support provided by the lower channel to the higher channel, but also the control the order given from the higher channel to the lower channel. Figure 1-4-1 shows the mapping relation by LTE downlink channel.

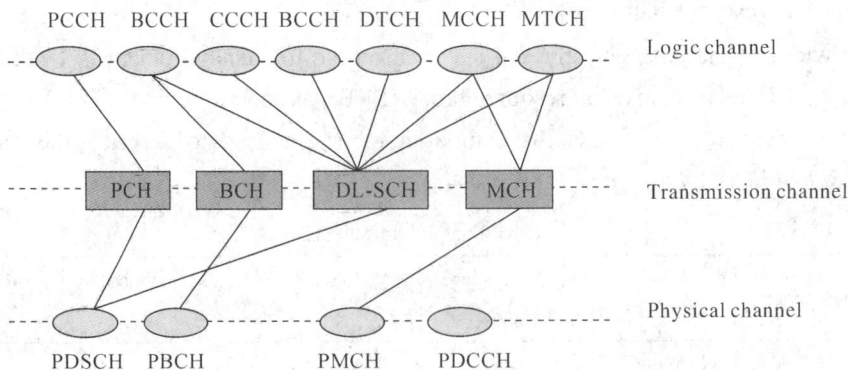

Figure 1-4-1　LTE downlink channel mapping

2. Downlink physical channel

The downlink physical signal includes the reference signal and the synchronizing signal.

In fact, the downlink reference signal is a pseudo-random sequence, and it has no information. The random sequence, which is composed of the time and frequency resource RE, is sent out. Therefore, the receiving terminal can evaluate the channel conveniently, and it can also provide the signal demodulation reference for the receiving terminal. The downlink reference signal includes:

(1) the special area reference signal related to the non-MBSFN transmission;

(2) MBSFN reference signal related to MBSFN transmission;

(3) the special reference signal for terminal.

The downlink synchronizing signal can synchronize the time and frequency of UE and eUTRAN during area searching. Synchronizing the time and frequency is the necessary condition for the UE and eUTRAN to work. There are two downlink signals, and they are as follows:

(1) main synchronizing signal;

(2) auxiliary synchronizing signal.

3. Uplink physical channel

Uplink transmission is similar to downlink transmission. The physical channel and the reference channel can be used to transmit information when UE needs to give information to eNB. LTE has three uplink physical channels, and they are as follows:

(1) Physical Uplink Shared Channel, PUSCH;

(2) Physical Uplink Control Channel, PUCCH;

(3) Physical Random Access Channel, PRACH.

Figure 1-4-2 shows the reflection relation of uplink channels.

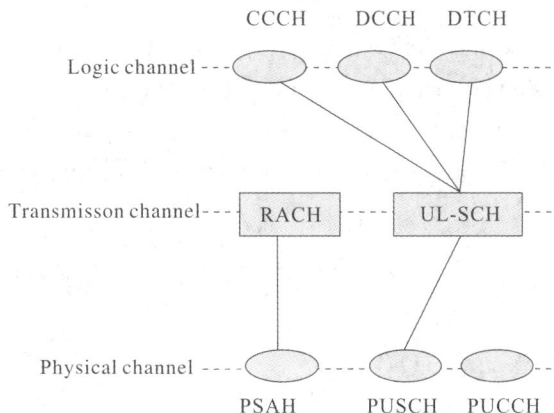

Figure 1-4-2   LTE uplink channel reflection relation

4. Uplink physical signal

The uplink physical signal includes:

(1) Dedicated Reference Signal, DRS;

(2) Sounding Reference Signal, SRS.

Work Task 5:   LTE Multi-antenna Technology

1. Transmission diversity

The transmission diversity contains cyclic time delay diversity, time-space/frequency code, and antenna switching diversity.

1) Cyclic time delay diversity

The time delay diversity means that the different time delay copies of the same signal are transmitted by different antennas. For the signal, the wireless channel can be enlarged through a time delay extension, which is to increase the frequency choice of the channel. The transmission antenna diversity can be transformed into frequency diversity. Figure 1-5-1 shows the time delay diversity for a double-antenna.

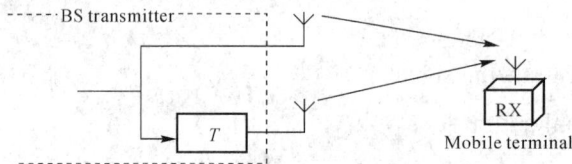

Figure 1-5-1   Time delay diversity diagram

In fact, one terminal can only receive one sent signal when time delay diversity is used; all the time delay diversity can be easily used in any wireless communication system without standard support. It is necessary for reference signal to use CDD (Cyclic Delay Diversity) to evaluate the equivalent space channel. Thus, a higher requirement is made that the reference signal should evaluate a larger extension of time delay. Normally, the time delay diversity can delay smaller time delay when it is used.

It is convenient to use the cyclic time delay diversity for OFDM transmission. Before increasing CP, the time domain sample value of the signal sent by different antennas can be cycled and shifted, so the frequency diversity can be increased. For OFDM, the circular shift of time domain signal should be correlated to the phase deviation of frequency domain, as shown in Figure 1-5-2.

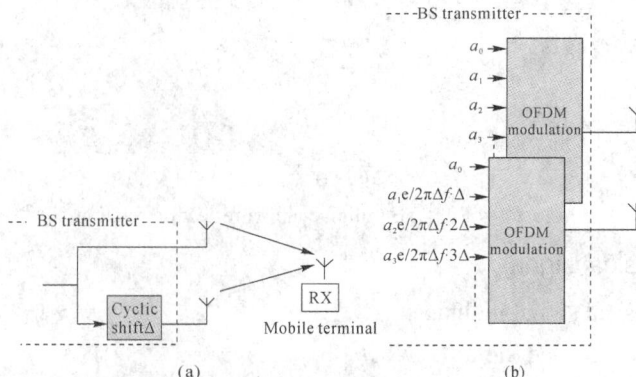

Figure 1-5-2   Cyclic time delay diversity diagram

2) Time/Space code

The most famous Space Time Block Code (STBC) method for two transmitting antennas is Alamouti. The first antenna can transmit the original signal, and for the second antenna, the transmission sequence of the signal can be changed by using two codes as a group, to conjugate and/or do the opposite operation, as shown in Figure 1-5-3.

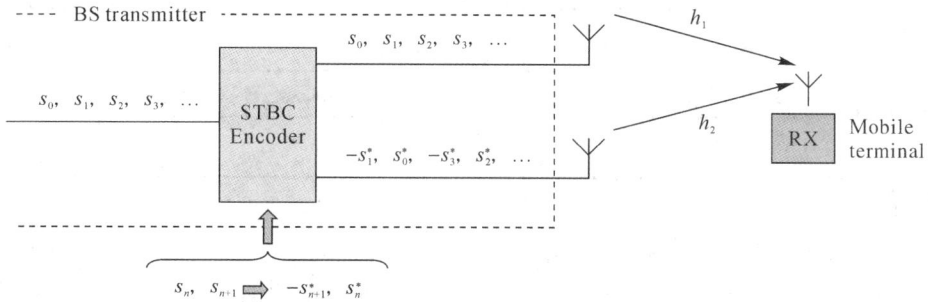

Figure 1-5-3   STBC diagram

If the above code is not related to the different subcarrier code or the code in the time domain, but the code Space Frequency Block Code (SFBC), as shown in Figure 1-5-4.

3) Antenna switching diversity

The antenna switching diversity technology means that an antenna is chosen one after another

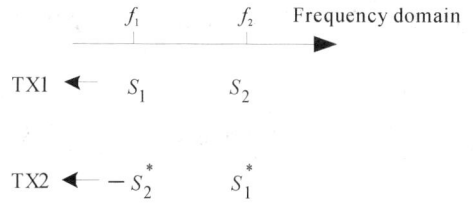

Figure 1-5-4   SFBC diagram

according to time and frequency to transmit data when there are many antennas on the sending terminal. If the antenna switching happens according to time, it is Time Switched Transmit Diversity (TSTD); if it happens among different subcarriers, it is Frequency Switched Transmit Diversity (FSTD), as shown in Figure 1-5-5 and Figure 1-5-6.

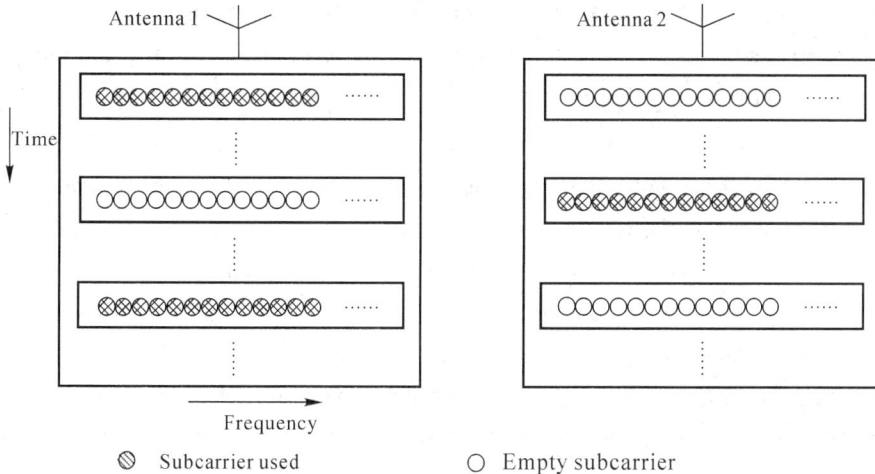

Figure 1-5-5   Antenna switching diversity TSTD

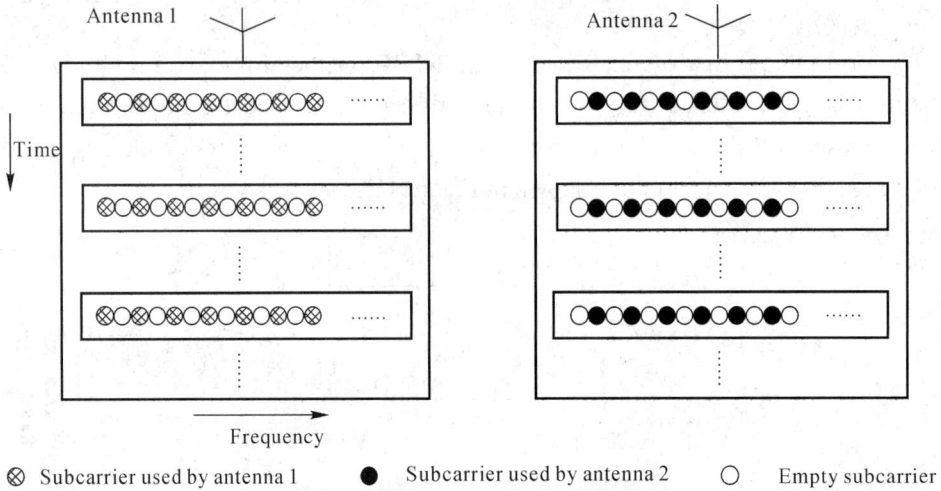

⊗　Subcarrier used by antenna 1　　　●　Subcarrier used by antenna 2　　　○　Empty subcarrier

Figure 1-5-6　　Antenna switching diversity FSTD

FSTD can be connected with SFBC easily, and it supports the transmission diversity technology for 4 transmitting antennas. Figure 1-5-7 shows the SFBC and FSTD combined transmission diversity which is supported by LTE.

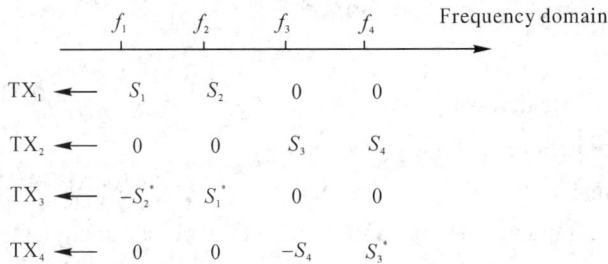

$$
\begin{array}{ccccc}
 & f_1 & f_2 & f_3 & f_4 \\
TX_1 \leftarrow & S_1 & S_2 & 0 & 0 \\
TX_2 \leftarrow & 0 & 0 & S_3 & S_4 \\
TX_3 \leftarrow & -S_2^* & S_1^* & 0 & 0 \\
TX_4 \leftarrow & 0 & 0 & -S_4 & S_3^*
\end{array}
$$

Figure 1-5-7　　SFBC+FSTD

2. Beam-forming

Based on the relation of antennas which use beam-forming, the beam-forming can be divided into two categories; they are the traditional long-term beam-forming and the short-term beam-forming, and the latter is based on precoding.

1) Long-term beam-forming

When the antenna is highly related to each other and the normal array antennas have small space, the long-term beam-forming can be used, as shown in Figure 1-5-8(a). The same signal can use different phase deviation, and it can be reflected on different antennas to be transmitted. Because of the high correlation among antennas, a special and large designated beam can be formed at the transmitting terminal, as shown in Figure 1-5-8(b). By adjusting the phase deviation used by the different antennas, the beam direction can be changed so as to intensify the signal and reduce the interference to other directions. The phase value can be obtained by estimating the wave direction of the signal.

Fraction of a wave length

Signal to be transmitted

(a)　　　　　　　　　　　　　　　(b)

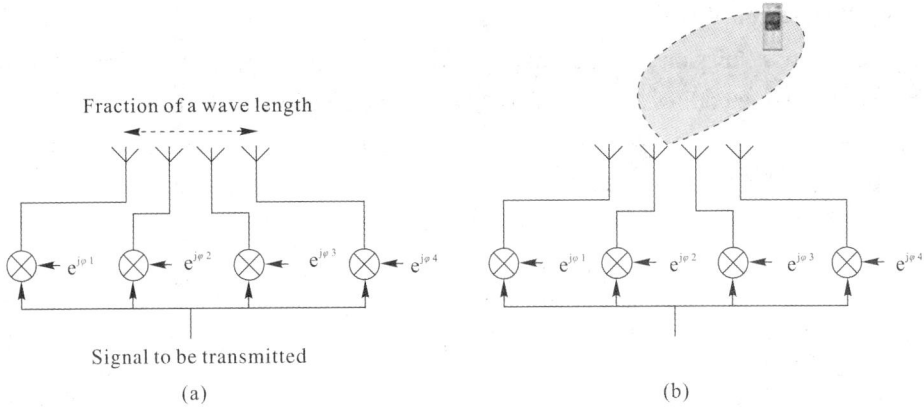

Figure 1-5-8　Traditional long-term beam-forming diagram

2) Beam-forming based on the precoding

If the correlation between antennas is small, the normal antenna array is the large spacing antenna array, or the polarization antenna array through which the signal is transmitted in different polarization directions. Compared with long-term beam-forming, beam-forming based on precoding also uses different transmission values on different antennas when the antenna correlation is small. The difference is that the transmission value here contains not only the phase adjustment but also the range adjustment, as shown in Figure 1-5-9.

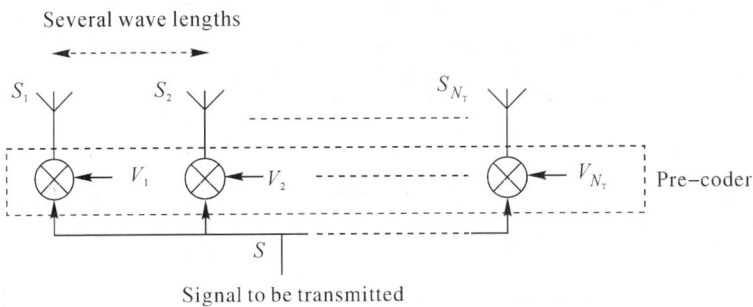

Several wave lengths

S_1　S_2　S_{N_T}

V_1　V_2　V_{N_T}　Pre−coder

S

Signal to be transmitted

Figure 1-5-9　Beam-forming based on precoding

Compared with long-term beam-forming, beam-forming based on precoding needs more detailed channel information to calculate the assigned value, for example, the instant channel lose. The update of the assigned value should be finished in a short time so as to obtain the instant lose change. Therefore, beam-forming based on precoding can provide not only the assigned increase but also the diversity increase.

3. Spatial reuse forming value

The spatial reuse of LTE has the following characteristics:

- Support Multi-code transmission.
- Use precoding technology, namely the spatial reuse of the closed cycle.
- Used together with CDD.
- Support MU-MIMO (Multi-User Multiple-Input Multiple-Output).

1) Multi-code transmission

Multi-code transmission means that the data stream mapped to antennas can have independent channel encoding and modulation; and single code transmission means that a data stream can be mapped to several antennas after the channel encoding and modulation, as shown in Figure 1-5-10.

The largest code number supported by LTE is 2.

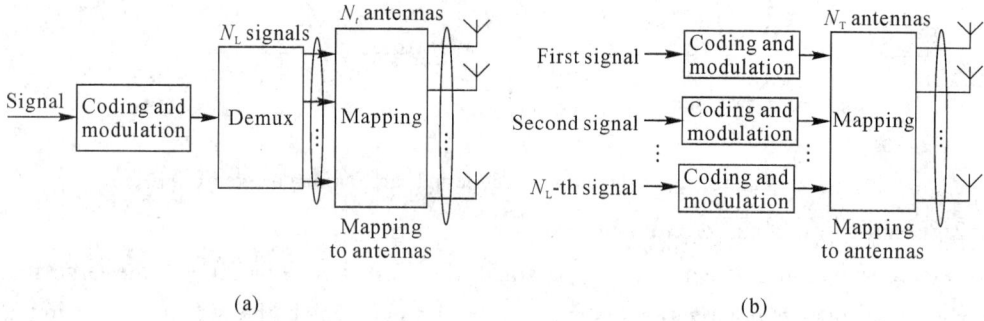

(a)   (b)

Figure 1-5-10   Single code and multi-code diagram

2) Precoding technology

Similar to beam-forming based on precoding, the spatial reuse based on precoding means that an precoding array is used to linear weight several data streams before their transmission, as shown in Figure 1-5-11.

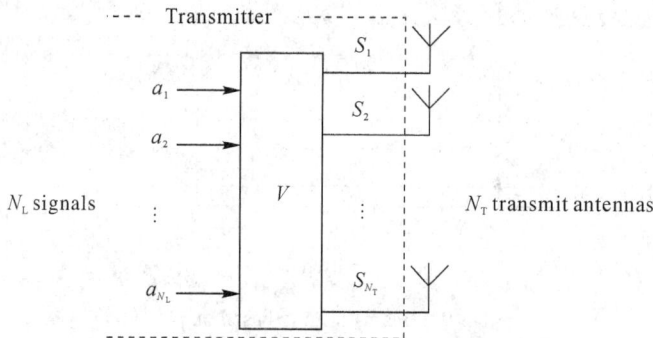

Figure 1-5-11   Spatial reuse based on precoding

There are two main purposes for the spatial reuse based on precoding:

(1) If the number of spatial reuse is equal to the number of transmitting antennas, the precoding can be used to orthogonalize several parallel transmissions so as to increase the signal isolation at the receiving terminal.

(2) If the number of spatial reuse used is smaller than the number of transmitting antennas, the precoding can support N spatial reuse signals to reflect on N antennas, and it can increase spatial reuse and beam-forming result.

3) Connected to CDD

The transmission signal will finish the CDD operation first and then the precoding operation if CDD time delay is large, as shown in Figure 1-5-12 (just take 2 antennas as an example).

$$x = WDUa \qquad U = \begin{bmatrix} 1 & 1 \\ 1 & -1 \end{bmatrix} \qquad D = \begin{bmatrix} 1 & 0 \\ 0 & e^{-j2\pi(i)/2} \end{bmatrix}$$

Sequence of the signal

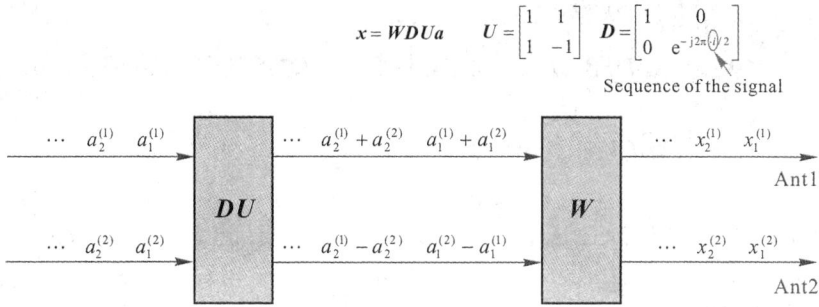

Figure 1-5-12 Connection mode for spatial reuse and large time delay CDD

4) MU-MIMO

When the base station sends several data streams that occupy the same time frequency resource to a user, it is called the single user MIMO (SU-MIMO) or space division multiplexing (SDM). When the base station sends several data streams that occupy the same time frequency resource to different users, it is called the multi-user MIMO (MU-MIMO) or space division multiple access (SDMA), as shown in Figure 1-5-13.

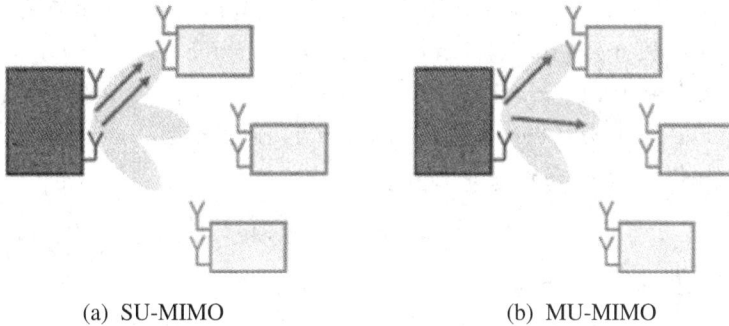

(a) SU-MIMO   (b) MU-MIMO

Figure 1-5-13 Downlink MIMO

Different from the downlink multi-user MIMO, the uplink multi-user is a virtual MIMO system. Each terminal will send a data stream, but two or more data streams use the same time frequency resource. For the receiver, these data streams sent by different terminals can be treated as the data streams sent by different antennas of the same terminal, thus a MIMO system is formed. As shown in Figure 1-5-14, (a) is the traditional MIMO system, which is the single user MIMO (SU-MIMO); and (b) is multi-user MIMO (MU-MIMO).

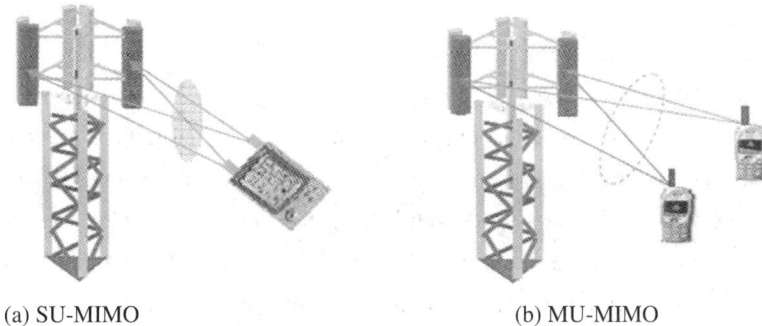

(a) SU-MIMO   (b) MU-MIMO

Figure 1-5-14 Uplink MIMO

Work Task 6:　Channel Adaptation and Channel-dependent Scheduling

One of the typical characteristics of wireless mobile communication channel is the rapid and large-range change of the instant channel, channel-dependent scheduling and link adaptation can make full use of the characteristic so as to improve the transmission quality of the wireless link.

Channel-dependent scheduling and link adaptation are normally used together.

1. Link adaptation technology

There are two link adaptation technologies, which are power control technology and speed control technology.

1) Power control technology

The purpose for power control is to dynamically adjust the TX power to maintain a certain level of signal to noise ratio at the receiving terminal, so that the transmission quality of the link can be guaranteed. The TX power should be increased if the channel condition is poor, and the TX power should be reduced if the channel condition is good. Therefore, the stable transmission rate can be sustained, as shown in Figure 1-6-1.

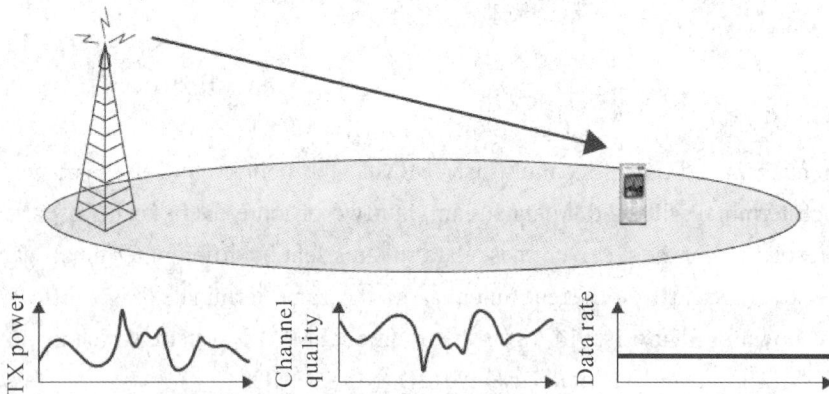

Figure 1-6-1　Power control diagram

2) Speed control technology

If the transmitting power is constant, the speed control technology is to guarantee the link transmission quality by adjusting the modulation mode and encoding speed of the wireless link transmission. A smaller modulation mode and lower encoding speed can be chosen if the channel condition is poor, and a larger modulation mode can be chosen if the channel condition is good, so the transmission speed can be maximized, as shown in Figure 1-6-2.

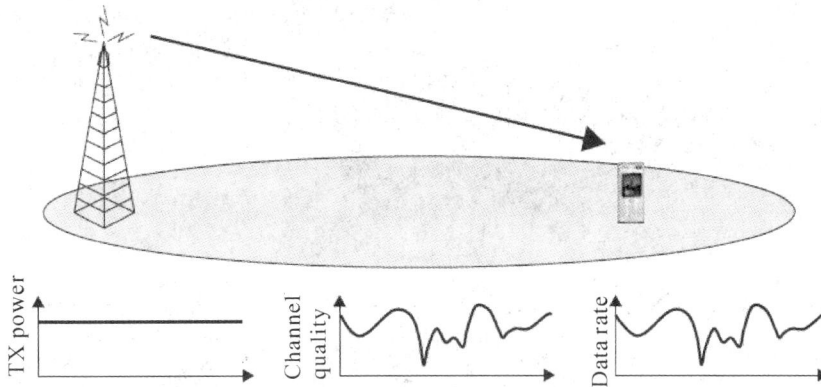

Figure 1-6-2   Speed control diagram

The efficiency of speed control should be higher than that of power control, this is because the full power can be used for transmission by speed control technology, while the power cannot be fully used by power control technology.

It doesn't mean that power control is not necessary, and the power control can greatly reduce the interference between users in a small area when non-orthogonal multi-address method (for example, CDMA) is used.

2. Channel-dependent scheduling

For the same resource block, the different users are in different positions in the mobile communication system so their signal transmission channels are different, too. The basic meaning of channel-dependent scheduling is that the user who has the best transmission condition should be scheduled for the related resource block so as to maximize the throughout capacity of the base station. The scheduling in this way is called the max-C/I scheduling, as shown in Figure 1-6-3.

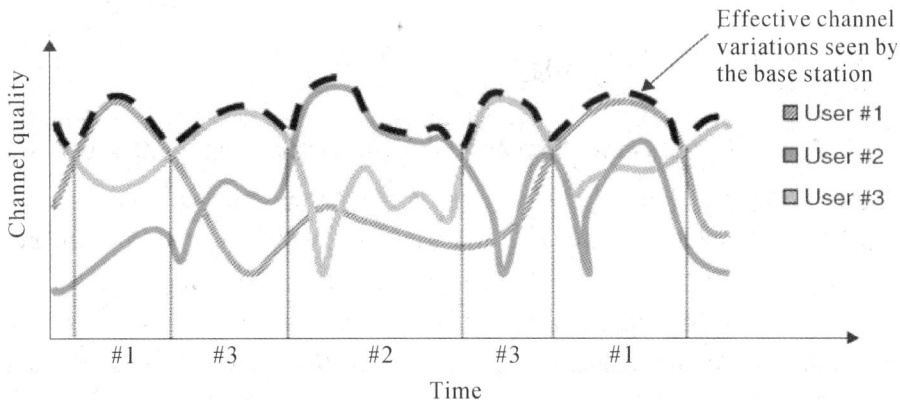

Figure 1-6-3   Channel-dependent scheduling diagram

For the single carrier CDMA system, the typical characteristics of the LTE system are channel scheduling and speed rate control in the frequency domain, as shown in Figure 1-6-4.

Time–frequency fading .
user #1

Time–frequency fading .
user #2

(a)

User#1 scheduled
User#2 scheduled
1ms

Time

180 kHz

Frequency

(b)

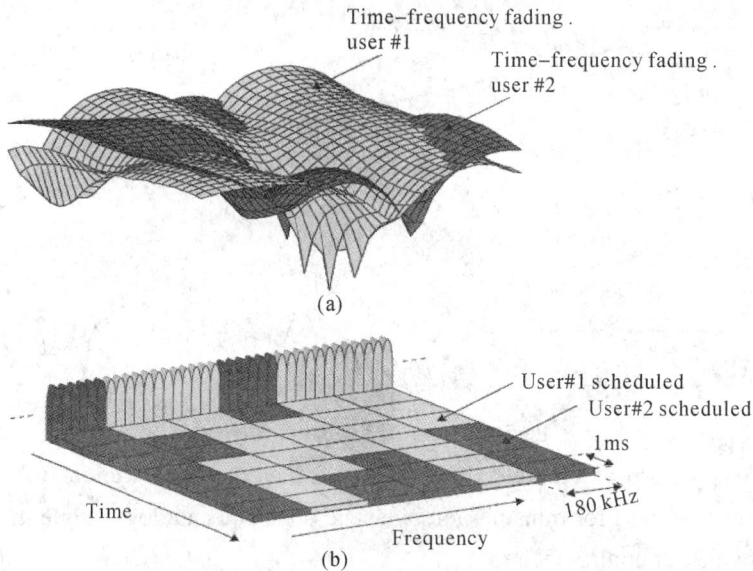

Figure 1-6-4　Frequency domain scheduling diagram

Therefore, the base station should know the channel information of the different frequency bands in the frequency domain. For the downlink, it can measure the public reference signal of the whole band to obtain the channel information of the different frequency bands, and these can be changed into a channel quality indicator (CQI) and given to base station. For the uplink, it can measure the uplink SRS transmitted by the terminal to obtain the channel information of the different frequency bands, and then to adjust the channel-dependent scheduling and speed rate control in the frequency domain.

Work Task 7: HARQ (Hybrid Automatic Repeat reQuest) Technology

1. FEC (Forward Error Correction) and ARQ (Automatic Repeat reQuest)

The fast fading characteristic of the wireless channel can be used to realise the channel-dependent scheduling and speed control, and there will always be some unpredictable interference which can cause the signal transmission failure. Therefore, the forward error correction (FEC) technology should be used. The working principle of FEC is to increase redundancy during signal transmission, which means the parity bits are inserted in the information bits before signal transmission. The parity bits use the method decided by encoding structure to calculate the information bits. Therefore, the real bits transmitted in the channel are much more than the original information bits, so the redundancy is introduced in transmitting signal.

Another method to solve the transmission error is to use the automatic repeat request (ARQ) technology. In ARQ method, the receiving terminal can evaluate whether the data packet received is right or wrong, and also it can send ACK (ACKnowledge) character to the transmitter to confirm the data. If the data packet is judged to be false, it can send NACK (Negative ACKnowledge) character to the transmitter which can send the same information back again.

Most systems use FEC and ARQ together, and it's called hybrid automatic repeat request, Hybrid ARQ (HARQ). HARQ can use FEC to correct all the false parts and can evaluate the uncorrectable mistakes through mistake detecting. The false data packet received will be abandoned, and the receiver will ask to send the same data packet back again.

2. Uplink and downlink HARQ of LTE

LTE use several parallel stop-and-wait HARQ protocols. The stop-and-wait means that it cannot transmit any data through the same process after HARQ process is used to transmit data packet and before any feedback information is received. The single stop-and-wait protocol is much simpler, but the transmission efficiency is low; but the low transmission efficiency can be improved by using several parallel stop-and-wait protocols and starting several HARQ processes. The basic working principle is that several HARQ processes are used, and the free processes can be used to transmit data packet when waiting for the feedback information of a certain HARQ process. The parallel processes need to ensure that any transmitter can use one of the processes to transmit information in RTT (Round Trip Time). Figure 1-7-1 shows the downlink transmission of FDD(Frequency-division Duplex). RTT includes the downlink signal transmission time T_P, downlink signal receiving time T_{sf}, downlink signal processing time T_{RX}, uplink ACD/NACK transmission time T_P, uplink ACK/NACK receiving time T_{TX}, and uplink ACK/NACK processing time T_{RX}, which is RTT $= 2*T_P + 2*T_{sf} + T_{RX} + T_{TX}$. The number of process is equal to the number of downlink subframe in RTT, which is $N_{proc} = \text{RTT}/T_{sf}$. It can be found out that RTT $= 2*T_P$ when the receiving and processing time of a signal is not taken into consideration, which is also the total time for a signal to be transmitted back and forth.

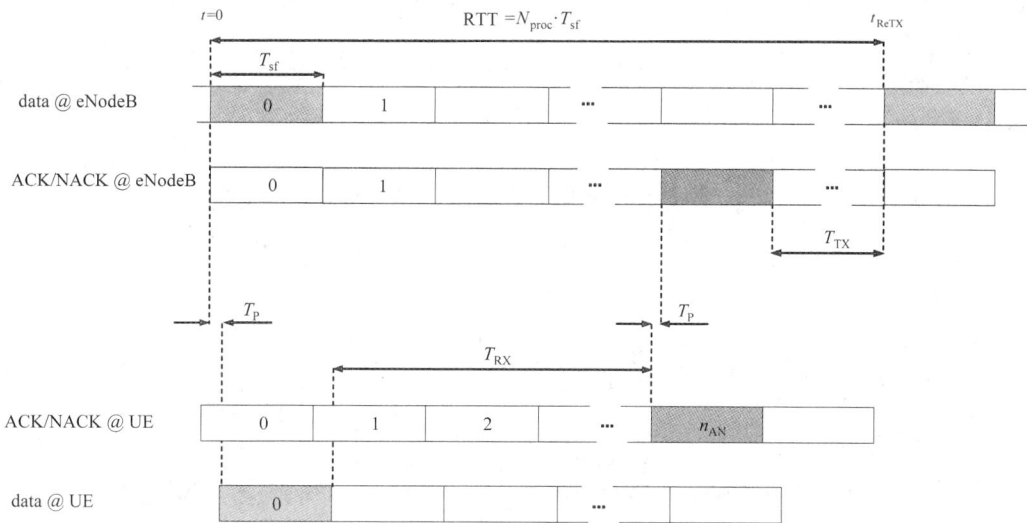

Figure 1-7-1 Downlink HARQ RTT and process number (FDD)

For the uplink HARQ of FDD, RTT = $2*T_P + 2*T_{sf} + T_{RX} + T_{TX}$, and the process number is $N_{proc} = RTT/T_{sf}$, as shown in Figure 1-7-2.

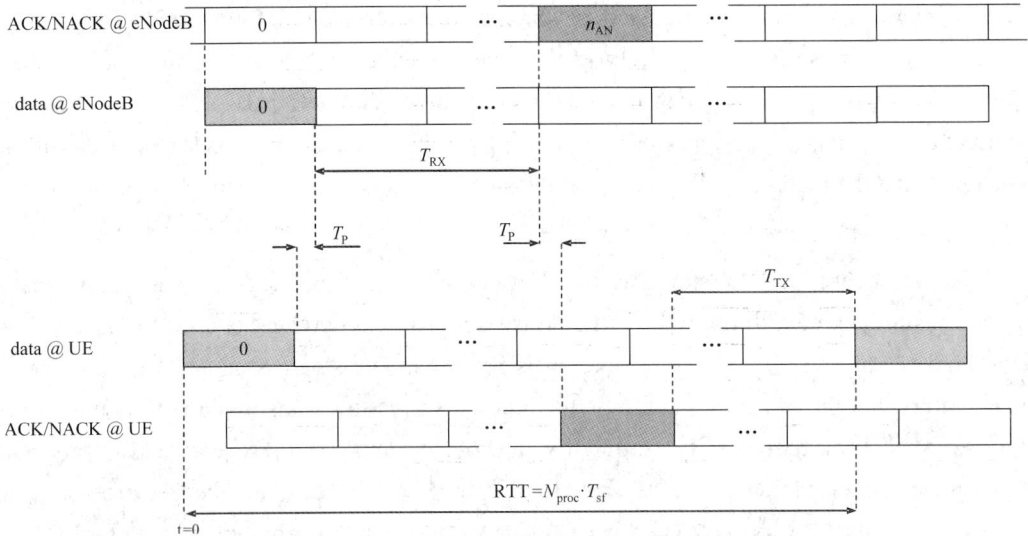

Figure 1-7-2　Uplink HARQ RTT and process number (FDD)

For TDD, the RTT size is related not only to transmission time delay, receiving time and processing time, but also to the TDD time slot ratio and the transmission location on the subframe. The working number is the number of subcarriers which have the same direction in RTT. Just take HARQ as an example, it can be supposed that the base station processing time is $3*T_{sf}$, the terminal processing time is $3*T_{sf} - 2*T_P$, the RTT and the process number of the different time slot ratio are different if it begins to transmit data to subcarrier 0, as shown in Figure 1-7-3.

At the same time slot ratio, the RTT and the process number are also different when the data transmission begins at the different subcarriers, as shown in Figures 1-7-4 and 1-7-5.

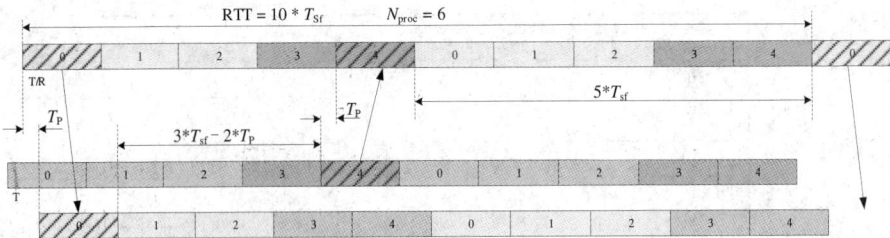

Figure 1-7-3　Subcarrier 0, DL: UL = 3:2

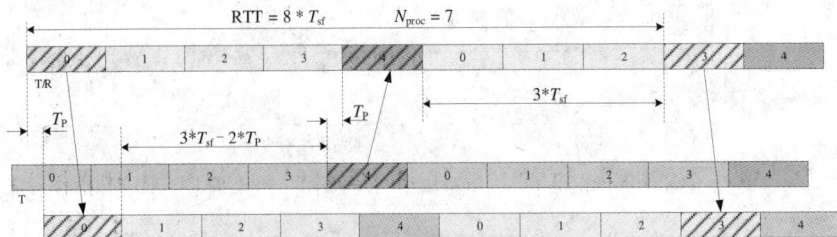

Figure 1-7-4　Subcarrier 0, DL: UL = 4:1

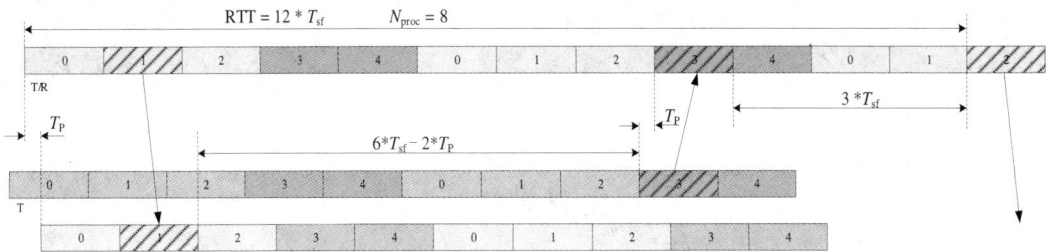

Figure 1-7-5　Subcarrier 1，DL: UL = 3:2

For one transmission, the response information (ACK/NACK) should be transmitted in the appointed time. For FDD, the transmission in any direction is continuous. The response can have a specified time interval for any transmission in any subcarrier. For example, LTE is sure to transmit information in any downlink subcarrier n, the ACK/NACK can be transmitted on the uplink subcarrier $n + 4$. For TDD, the transmission in any direction is not continuous, so the constant time interval for ACK/NACK and the previous transmission cannot be guaranteed. Thus, for TDD, the time set of ACK/NACK is related to time slot ratio and subcarrier location. If the processing time for the base station is $3*T_{sf}$ and the processing time for the terminal is $3*T_{sf} - 2*T_P$, the ACK/NACK transmission time under FDD and TDD mode is shown in Figures 1-7-6, 1-7-7 and 1-7-8.

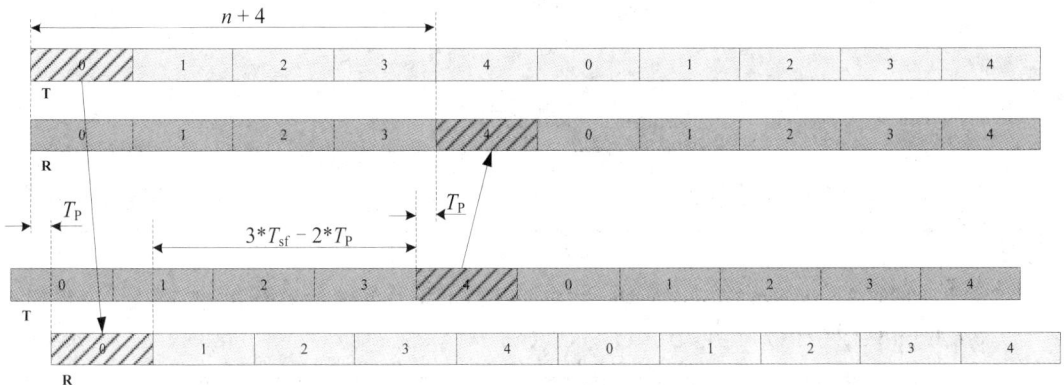

Figure 1-7-6　FDD

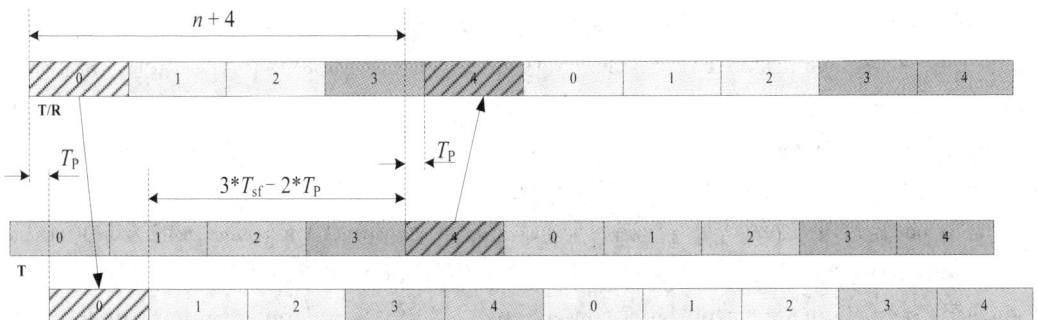

Figure 1-7-7　TDD, DL:UL = 3:2, subcarrier 0

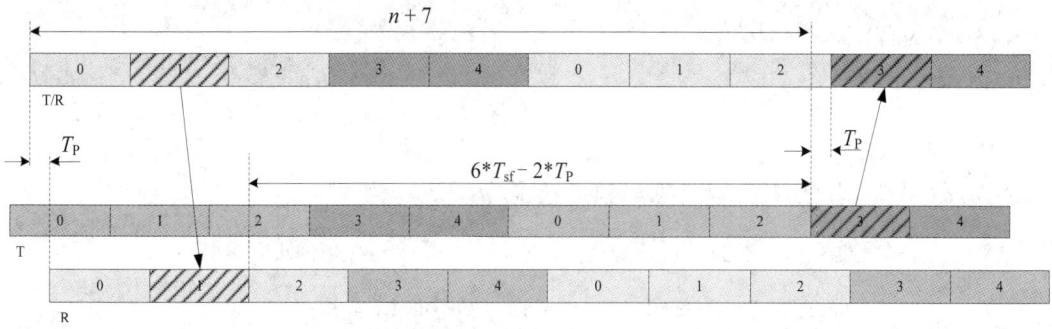

Figure 1-7-8   TDD, DL:UL = 3:2, subcarrier 1

If the retransmission is done in the predefined time, the receiver is not necessary to display the process number, which is called the synchronous HARQ protocol. If the retransmission is done in any useable time after the previous transmission, the receiver is necessary to display the exact process number, which is called the asynchronous HARQ protocol, as shown in Figure 1-7-9.

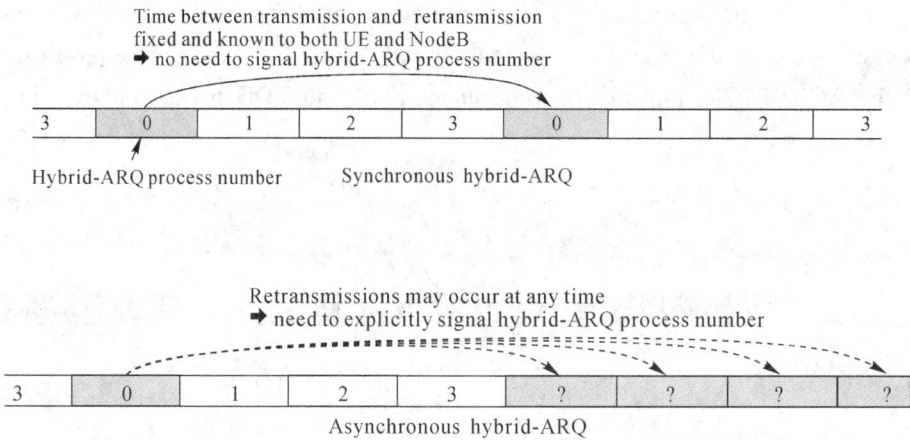

Figure 1-7-9   Synchronous HARQ and asynchronous HARQ

Synchronous HARQ doesn't mean the primary transmission and retransmission have the fixed time interval, and this should be known in advance. Different time slot ratio can choose different RTT to reduce the uplink transmission time delay.

LTE downlink uses asynchronous HARQ, and uplink uses synchronous HARQ.

HARQ can be adaptive and non-adaptive. The adaptive HARQ means the part or the whole of the primary transmission can be changed during retransmission, for example, modulation mode, resource distribution, etc. Additional signal orders are needed to notify the changes. The non-adaptive HARQ means that the retransmission change is decided by the transmistter and the receiver, and it doesn't need additional signal orders.

LTE downlink uses the adaptive HARQ, and uplink supports both adaptive and non-adaptive HARQ. Only the NACK response information in PHICH(Physical Hybrid ARQ Indicator Channel) channel can trigger non-adaptive HARQ; the adaptive HARQ can be realized by PDCCH(Physical Downlink Control Channel) scheduling, which means the base station doesn't give NACK when it finds the false receiving or transmitting, and the retransmission parameter is scheduled by the dispatcher.

Work Task 8:   Cell Searching

1. Time frequency resource

LTE system uses the main synchronous signal and secondary synchronous signal for cell searching. For the subframe 1, the main/secondary signal is reflected on the middle 72 subcarriers of the last/second from the back OFDM in the time slot 1 and time slot 10, as shown in Figure 1-8-1 (take normal CP as an example).

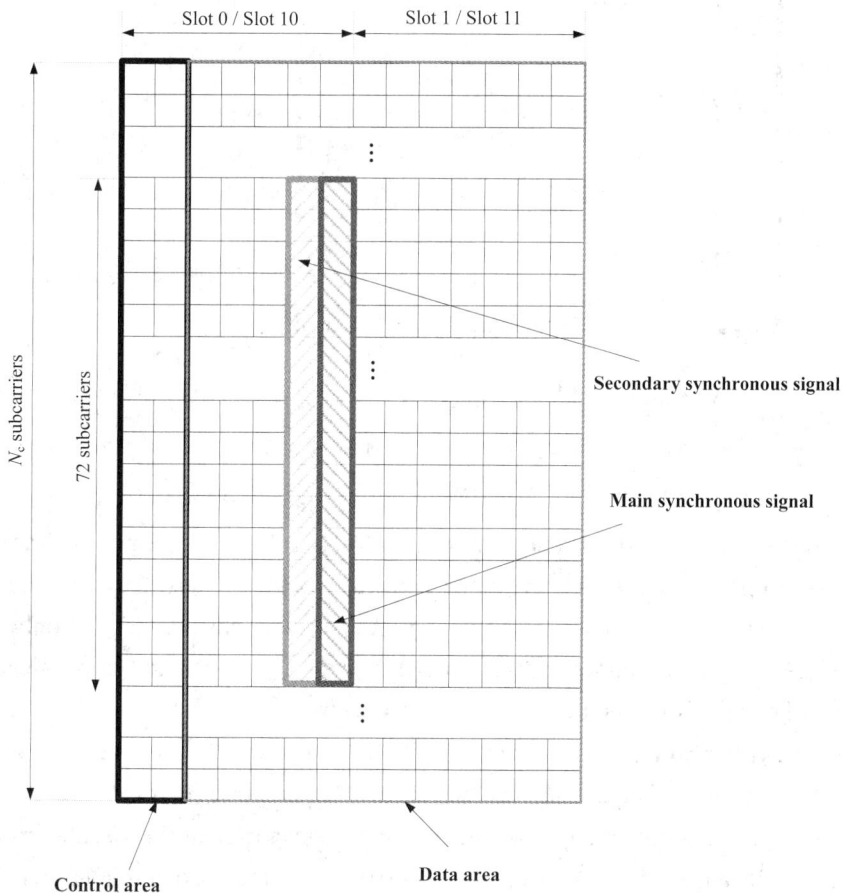

Figure 1-8-1   Main/secondary synchronous signal location diagram (frame 1, normal CP)

For frame structure 2, the main synchronous signal is reflected on the middle 72 subcarriers of the first OFDM in DwPTS(Downlink Pilot Time Slot), and the secondary synchronous signal is reflected on the middle 72 subcarriers of the last OFDM in time slot 1 and time slot 11, as shown in Figure 1-8-2 (take normal CP as an example).

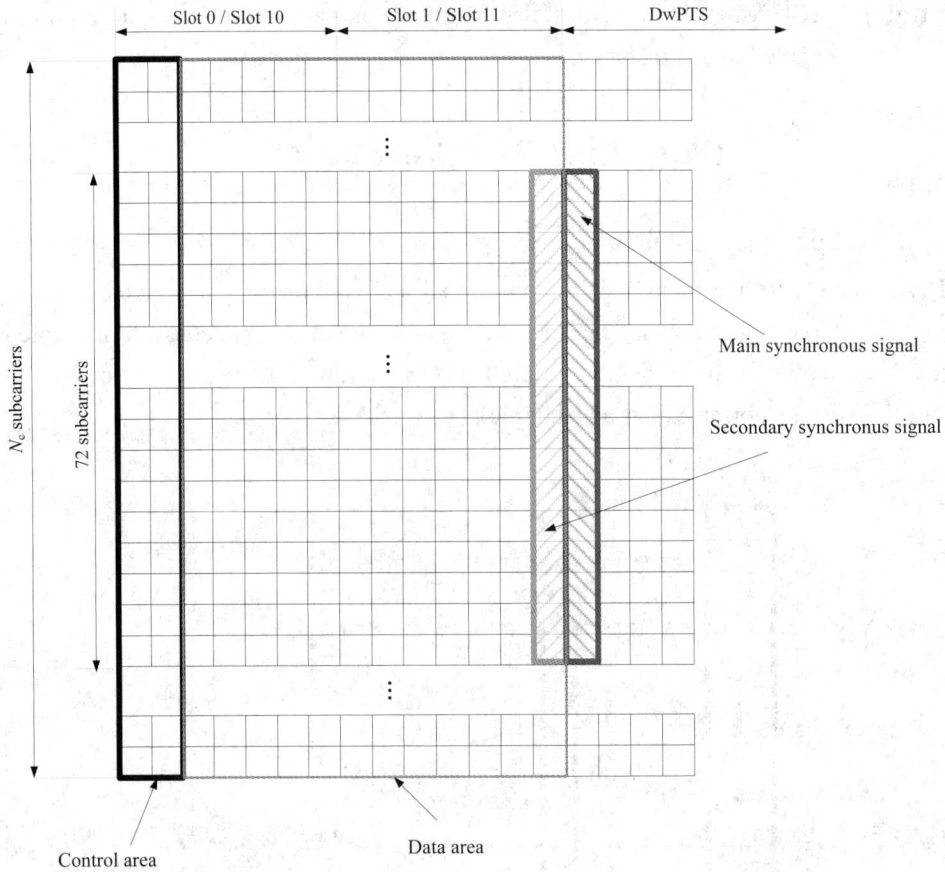

Figure 1-8-2　Main/secondary synchronous signal location diagram (frame structure 2, normal CP)

2. Sequence generation

The sequence used by synchronous signal is closely related to the physical area ID. LTE supports 504 physical area IDs, which can be divided into 168 groups; they are called physical area ID group and each group has 3 physical area IDs. Thus, a physical ID can only be defined by the physical ID group number $N_{ID}^{(1)}$ (the range is from 1 to 167) and the physical ID group number $N_{ID}^{(2)}$ (the range is from 0 to 2), that is $N_{ID}^{cell} = 3N_{ID}^{(1)} + N_{ID}^{(2)}$.

The main synchronous signal uses Zadoff-Chu sequence which has three main synchronous sequences, and the length is 62. They can be mapped on the three physical area $N_{ID}^{(2)}$, the main synchronous sequence is only mapped on the middle 62 subcarriers of the 72 subcarriers, and the 5 subcarriers at the edge are left as the protection space. The main synchronous signal transmits in 5 ms a time, the forward half-frame and the backward half-frame in a wireless frame use the same sequence.

The sequence used by the secondary synchronous signal results from the interweaving cascade of two binary sequences, each of which has a length of 31, and it is scrambled by the scrambling sequence decided by the main synchronous signal. The binary sequence with a length of 31 and the scrambled sequence are all produced by m sequence. There are 168 secondary synchronous sequence groups, each of which has a length of 62. Among these groups, the binary sequence with a length of 31 has a one to one mapping relation with the 168 physical area ID group $N_{\text{ID}}^{(1)}$. The scrambled sequence is decided by the physical group ID $N_{\text{ID}}^{(2)}$. Secondary synchronous sequence is only mapped on the middle 62 subcarriers of the 72 subcarriers, and the 5 subcarriers at the edge are preserved as the protection space.

One secondary synchronous sequence group has two secondary synchronous sequences with the length of 61 which are transmitted through the forward half-frame and the backward half-frame in a wireless frame, which means that the secondary synchronous signal transmits in 5ms each time. However, the forward half-frame and the backward half-frame in a wireless frame use different sequences.

3. Cell searching process

LET cell searching goes as follows:

(1) Obtain 5 ms time set and physical area ID from P-SCH (Primary Synchronization CHannel).

UE uses the basic synchronous sequence to confirm the code synchronization and identify the area ID in the area groups, which can be obtained by UE through contrasting the relation of the three local basic synchronous sequences and the signals received.

(2) Obtain the wireless frame time set and area group ID from S-SCH (Secondary Synchronous CHannel).

• UE uses the secondary synchronous sequence to confirm the wireless frame time set and area ID in the first step, which can be obtained by UE through contrasting the relativity of all secondary synchronous sequences and the signal received;

• UE can obtain the area ID through the detected area group ID index and the area ID in the local ID group;

• During the S-SCH detecting , the CP length can also be detected and obtained at the same time.

(3) Obtain the area ID verification from the downlink reference signal (optional).

(4) Read BCH (Broadcast Channel).

Work Task 9:   Random Access

1. Time frequency resource

The basic structure of the LTE(Long Term Evolution) system's physical random access channel (PRACH) Preamble is shown in Figure 1-9-1.

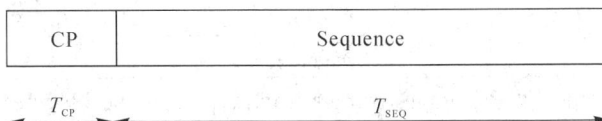

CP	Sequence
T_{CP}	T_{SEQ}

Figure 1-9-1   Preamble structure diagram

　　LTE defines 5 types of PRACH Preamble structures, which occupy 6 PRB (72 subcarriers) on the frequency domain, as shown in Table 1-9-1.

Table 1-9-1　　Preamble structure

Preamble format	Time length	T_{CP}	T_{SEQ}	Sequence length	GT
0	1 ms	$3152 \times T_s$	$24\,576 \times T_s$	839	$\approx 97.4\ \mu s$
1	2 ms	$21\,012 \times T_s$	$24\,576 \times T_s$	839	$\approx 516\ \mu s$
2	2 ms	$6224 \times T_s$	$2 \times 24\,576 \times T_s$	839 (transmit twice)	$\approx 197.4\ \mu s$
3	3 ms	$21\,012 \times T_s$	$2 \times 24\,576 \times T_s$	839 (transmit twice)	$\approx 716\ \mu s$
4(can only use FS2)	$\approx 157.3\ \mu s$	$448 \times T_s$	$4096 \times T_s$	139	$\approx 9.4\ \mu s$

　　Preamble format 4 can only be transmitted in the UpPTS domain of frame structure 2.

　　In order to support different numbers for PRACH, LTE has several types of PRACH settings. It defines the time and location of PRACH happening, and the high level can give the notice of the frequency domain location.

　　For FDD, PRACH settings are shown in Table 1-9-2, and one subframe can only transmit one PRACH.

Table 1-9-2　　FDD PRACH setting

PRACH setting	System frame number	Subframe number
0	even number	1
1	even number	4
2	even number	7
3	any one	1
4	any one	4
5	any one	7
6	any one	1, 6
7	any one	2 ,7
8	any one	3, 8
9	any one	1, 4, 7
10	any one	2, 5, 8
11	any one	3, 6, 9
12	any one	0, 2, 4, 6, 8
13	any one	1, 3, 5, 7, 9
14	any one	0, 1, 2, 3, 4, 5, 6, 7, 8, 9
15	even number	9

　　FDD supports 4 types of Preamble formats, each type can use 16 types of PRACH settings, and it needs 5 bits to indicate these information.

For TDD, the uplink subframe numbers are different under different subframe ratios, and it is probable that there is only one uplink subframe in 10 ms. Thus, TDD allows one uplink subframe to transmit several PRACH which should be distributed first on the time domain. If all of them cannot be put on the time domain, then several PRACHs are allowed to be put on the frequency domain. TDD supports five types of Preamble formats, and the Preamble format 4 can only be transmitted in UpPTS. Now, it is sure that the Preamble format indicator and PRACH setting indicator can be jointly encoded to form the extent PRACH setting. The exact PRACH setting is still not defined.

2. Sequence production

The LTE Preamble sequence is produced by Zadoff-Chu sequence, and one area supports 64 Preamble sequences. The area Preamble is produced by the Zadoff-Chu root sequence to conduct different circular shift. If this method cannot provide enough Preamble numbers, the logic sequence and the neighboring Zadoff-Chu root sequence can be used. The root (beginning) number information can be obtained by area broadcasting. Pramble formats 1-3 have 838 root sequences; Preamble format 4 has 138 root sequences.

3. Random access process

The following 5 states need the random access:

(1) primary access from LTE_IDLE;

(2) primary access after the wireless link failure;

(3) switching;

(4) the downlink data arrival under the LTE_ACTIVE condition, for example, there is no uplink synchronous condition (OUT_OF_SYNC);

(5) The uplink data arrival under the LTE_ACTIVE condition, (for example, there is no uplink synchronous condition (OUT_OF_SYNC), or there is no Scheduling Request (SR) sent by PUCCH(Physical Uplink Control Channel) resource).

The LTE random access process includes two types: the Contention based random access can be used in all 5 situations mentioned above, and the Non-contention based is only used for switching and the downlink data arrival.

Figure 1-9-2 shows the contention based random access and which is described as follows.

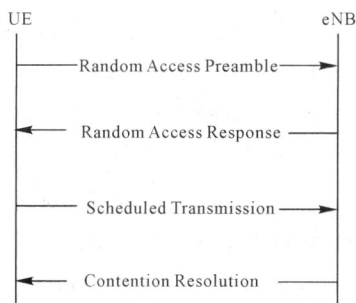

Figure 1-9-2   Contention based random access process

Step 1: UE randomly chooses a Preamble code and sends it through PRACH.

Step 2: Node B sends the downlink random access response after Preamble code is transmitted and detected, and the random access response has the following information:

① the number of Preamble code received;

② the related time adjustment information of Preamble code received;

③ the uplink resource location indicator distributed to the terminal;

④ the temporary C-RNTI (Cell-Radio Network Temporary Identifier) distribution.

Step 3: After receiving the random access response, UE will send the uplink information to the distributed uplink resource. The uplink information should at least have the terminal's unique ID (TMSI) or random ID.

Step 4: The base station uses DL-SCH to send the contention resolution information. By contrasting the information ID received in step 4 with the ID sent in step3, UE considers the access is successful if they are the same. If the users use the same Preamble in step 1 and have same interference, or the users are distributed the same temporary indicators in step2, only one of the users can be accessed successfully.

Non-contention random access process is shown in Figure 1-9-3 and described as follows.

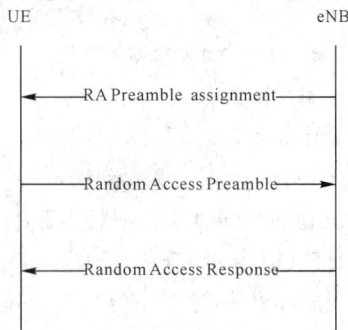

UE   eNB

←————RA Preamble assignment————

————Random Access Preamble————→

←————Random Access Response————

Figure 1-9-3   Non-contention random access process

(1) Random access Preamble is distributed to terminal through signal order specially used by downlink, and this Preamble should not be included in the Preamble assemblage used by the contention based random access.

(2) The terminal sends random access Preamble.

(3) The base station uses DL-SCH to send random access response.

The random access response information includes the adjustment information used for switching and the initial uplink permission at a set time, the adjustment information used for the downlink data arrival at a set time, and the Preamble serial number detected.

Random access response information uses the DL-SCH channel for transmission, PDCCH indicates the exact time frequency location (PDSCH), and also RA-RNTI (Random Access Radio Network Temporary Identifier) is distributed for PDCCH.

The random access response information cannot use HARQ.

The random access response information of several users can be transmitted in the same DL-SCH.

Training Module 2: TD-LTE Products (12 Class Hours)

【Basic Description】

The products of Datang mobile TD-LTE (Time Division-Long Term Evolution) can be mainly divided into three parts, and they are base station (eNB), core network (Evolved Packet Core network，EPC) and network management center (Operation and Maintenance Center，OMC). The project has designed 3 typical work tasks for the workers who just begin to do the network installation for the mobile communication system, the purpose is to help the beginners to recognize, experience and understand the three parts of the equipment. The project content includes the recognition of the basic capacities and the functions of the base station products, core network products and OMC network management products, etc. The learners can meet the basic requirements for the mobile network installation and adjustment after trainings.

【Training Elements】

1. Knowledge purpose

(1) Master the basic information of the mobile communication system products.

(2) Understand the technical characteristics of the mobile communication system products.

(3) Recognize the appearance of the mobile communication system products.

(4) Master the software and hardware structures of the mobile communication system products.

(5) Master the service and functions of the mobile communication system products.

(6) Master the typical setting examples of the mobile communication system products.

2. Skill purpose

(1) Master the basic composition of the mobile communication system products.

(2) Master the service and functions of all kinds of the mobile communication system products.

(3) Master the typical setting methods of the mobile communication system products.

【Training Requirements】

1. Preparation of tools, instrument and equipment

A set of mobile communication equipment.

2. Knowledge evaluation points

(1) composition of the mobile communication equipment system.

(2) Functions of the various composition parts.

(3) Performance and functions of the core function unit.

3. Skill evaluation points

(1) Have the ability to draw the structure chart of the mobile communication system.

(2) Master the performance and function of the main products.

(3) List the two typical indoor and outdoor setting examples.

Work Task 1:   Production of Base Station

A complete TD-LTE mobile communication network includes terminal equipment, base station equipment, core network equipment, network management center and transmission network equipment, etc. All these equipment supports various kinds of TD-LTE service, as shown in Figure 2-1-1. The base station equipment is the most important wireless equipment in the network.

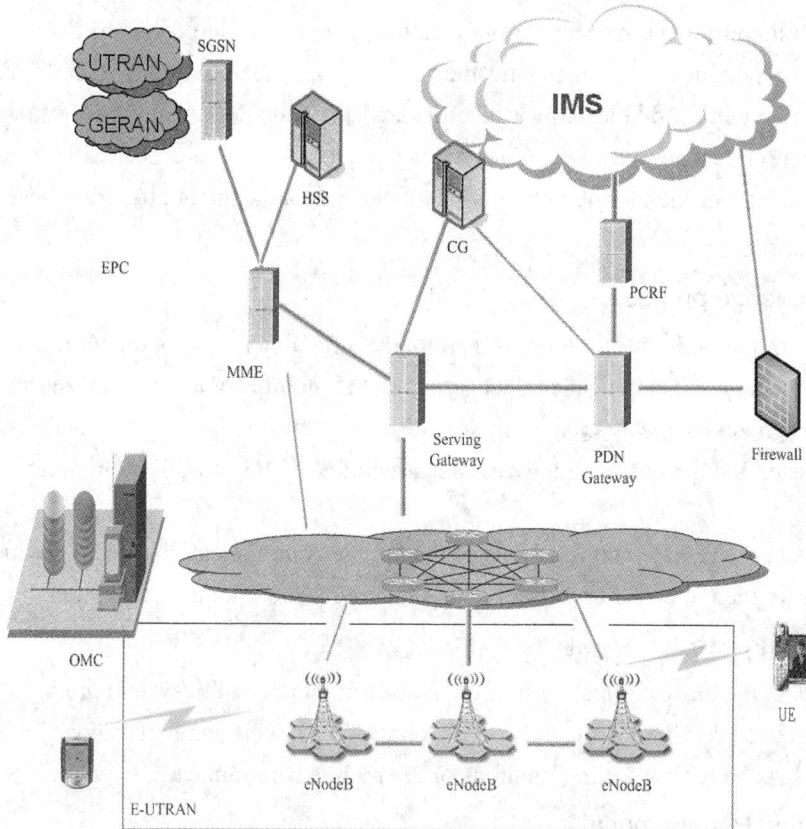

Figure 2-1-1   TD-LET network structure

1. Introduction to the base station products

Datang Mobile Communication Limited Company can provide a series of TD-LTE base stations, for example, the indoor base station and outdoor base station, and it can provide a solution to TD-LTE system wireless access, which is based on the different requirements of the users and working situations.

EMB5116 TD-LTE is a kind of baseband remote base station designed by Datang Mobile Communication Limited Company. By using baseband remote technology, it can cover not only the local area, but also the remote area, so as to solve the base address problem.

EMB5116 TD-LTE can be used for outdoor cover, for example, city downtown, suburbs, small towns, villages, special roads, etc. The use of remote technology can not only reduce cost, but also cover the target area quickly.

EMB5116 TD-LTE can also be used to solve the middle or small sized indoor cover, for example, tunnels, underground stations, tall buildings and communities, etc. It does not increase the costs, and can improve the net cover and service quality.

2. Technical characteristics of the base station products

1) Main technical characteristics

(1) Have advanced system structure for TD-LTE.

(2) Use the resource pool designing method to improve the hardware resource usage and system fault tolerance.

(3) Use the digital middle-frequency technology to improve the signal processing capacity;

(4) Have more powerful sector processing capacity and support large power cover and band width cover in single sector.

(5) Have the intelligent fan control capacity to improve the fan usage and reduce the noise.

(6) Support the self-adaptation filter and anti-interference in the band.

(7) Support RRU (Remote Radio Unit) to freely enlarge the wireless cover area.

2) Larger coverage

(1) The largest cover radius of the local area is 100 km.

(2) Support the intelligent antenna to improve the uplink receiving sensibility and increase the downlink cover area.

(3) Through optical fiber, the remote single-stage standard distance is 2 km, the largest distance for single-stage is 10 km, the largest distance for multi-stage is 40 km, and the highest stage is 4.

3) Flexible setting

(1) 20 M band width in each area can support 400 activated users, and the number of the connected users can be 1200.

(2) The standard number of area is 3, the highest processing capacity of band width is 60 M; it can support 1200 activated users and 3600 users to connect.

(3) Increase baseband board to increase the capacity.

(4) Support O1 and O2 area setting.

(5) Support S1/1/1 and S2/2/2 area setting.

4) Flexible network construction

(1) S1/X2 network interface. The single machine case supports 2, FE/GE self-adaptation

interfaces (the 2 interfaces can be distributed into electrical interface or optical interface respectively, and it supports an electrical one and an optical one, or the combination of them).

(2) Network construction method

① Support the same frequency network construction.

② Support S1/X2 star type, link type and circle type network construction.

③ Support Ir star type, link type and circle type network construction.

(3) Clock source. Support the synchronization of GPS, Beidou Satellite Navigation, GPS/ Beidou Satellite optical fibre remoteness, and the high-level eNode B.

5) Flexible installation

EMB5116 TD-LTE is small in size and light in weight, which can be conveniently installed in the indoor wall of buildings. The 19-inch standard equipment case can be directly installed in the indoor environment with no need for equipment room or air_conditioners. Therefore, the building of a base station can be fast and low in cost.

6) Convenient capacity update

(1) Compatible design. The board card of EMB5116 TD-LTE is compatible with all EMB-TD base stations of Datang Mobile Communication Equipment Limited Company.

(2) Flexible setting. Support all-direction area setting and multi-sector setting, and flexible cover can be realized easily.

(3) Convenient capacity extension. EMB5116 TD-LTE can increase the baseband board card to extent capacity, and the largest bandwidth of a single machine can be enlarged to 120 M if needed.

7) Powerful operation and maintenance function

(1) The mobile network management system provides the operation, maintenance and management function to LMT(Local Maintenance Terminal) (LMT is different from LMT-B of TD);

(2) Support system supervision, data setting, warning resolution, safety management, equipment operation, software setting, supervision management, self-setting optimization, tracking management, etc.

8) Serial products

EMB5116 TD-LTE supports the serial design of the related function units:

(1) Antenna serialization: support several kinds of all-direction, dual-polarization and sector array antennas like the indoor distributed antenna, 2-antenna and 8-antenna, etc.

(2) RRU power serialization: support multi-level amplifiers.

(3) Bandwidth serialization: each baseband board supports 10 M and 20 M bandwidth.

3. The appearance of the main equipment of the base station

The equipment size of EMB5116 TD-LTE is 88 mm × 483 mm × 310 mm(height × width × depth). The full weight is 10 kg, and the height of the equipment case is 2U. Figure 2-1-2 shows its appearance.

Figure 2-1-2　Outer appearance of equipment case

The equipment case has the following characteristics:

(1) Use the Aluminum alloy structure so the equipment case is not heavy.

(2) The whole case has electrical conductivity and its shielding effect is good.

(3) Have the excellent wind channel to guarantee good ventilation and heat dissipation.

(4) Installation and maintenance is easy.

(5) The appearance is simple and beautiful.

4. The hardware structure of the main equipment of the base station

EMB5116 TD-LTE main equipment includes Switch Control Transmission Board E Type (SCTE), baseband processing and Ir interface unit (Baseband Processing Only Board X Type, BPOX), and also includes Common Backplane (CBP), Fan Control Board (FC), Environment Monitor Board A/D Type (EMA/EMD) and Power Supply Board A/C Type (PSA/PSC), Extend Transmission Processing Board E Type (ETPE).

Figure 2-1-3 shows the hardware unit configuration, and Figure 2-1-4 has shown the full board.

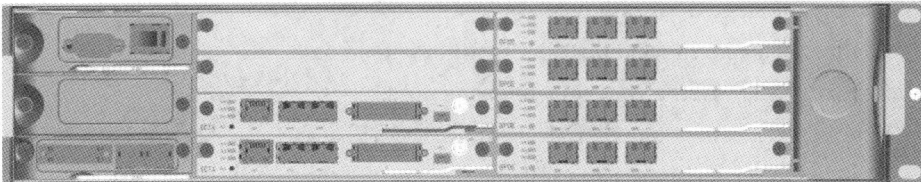

Figure 2-1-3　Hardware unit configuration in the main unit

SLOT 11	BPOG	SLOT 3	BPOG	SLOT 7	
PSA	BPOG	SLOT 2	BPOG	SLOT 6	
SLOT 10	SCTE	SLOT 1	BPOG	SLOT 5	FC
EMA SLOT 9		SLOT 0	BPOG	SLOT 4	SLOT 8

(a) Direct-current full board diagram

PSA SLOT 11	BPOG	SLOT 3	BPOG	SLOT 7	
	BPOG	SLOT 2	BPOG	SLOT 6	
SLOT 10	SCTE	SLOT 1	BPOG	SLOT 5	FC
EMA SLOT 9		SLOT 0	BPOG	SLOT 4	SLOT 8

(b) Alternating current full board diagram

Figure 2-1-4　Board diagram

1) Switch control and transmission unit

Switch control and transmission unit is composed of SCTE single board.

•The S1/X2 interface between EMB5116 TD-LTE and EPC supports 2 interfaces (electrical interface/optical interface), FE (FastEthernet)/GE (GigabitEthernet) self-adaptation interfaces in the full installation of single case (the 2 interfaces can be set as the electrical interface or optical interface respectively, and it supports the case that one is electrical and the other is optical or the combination of them).

The SCTE single board has the following functions:

(1) Function of service and signal orders switching;

(2) Maintenance for all control and uplink interface protocols;

(3) High reliable clock and its maintaining function;

(4) Power-on and power-saving control for a single board;

(5) On-site detection and survivor detection for a single board;

(6) Clock and synchronous code stream distribution;

(7) Equipment case management in dependent of single board software;

(8) Main system backup and redundant backup.

2) Baseband processing unit

BPOG (Baseband Processing Only Board G Type, baseband processing and Ir interface unit) single board has the following main functions:

(1) Realize the standard Ir interface.

(2) Realize the gathering and distributing of baseband data.

(3) Realize the TD-LTE physical layer calculation.

(4) Realize the L2 function of TD-LTE MAC/RLC/PDCP .

(5) Realize the self-operation and self-maintenance of the board.

3) Environmental monitoring unit

The environmental monitoring unit EMA/EMD provides the environment supervision to the base station, and the input/output interface for the intelligent point and the dry contact point. It supports the synchronous level linker of the base station, too.

EMA single board has the following main functions:

(1) Realize the outer environment supervision, dry contact point input/output and intelligent interface;

(2) Realize the outer clock link;

(3) Receive SCTE power control signal to control the up and down power, so as to realize the board electricity saving;

(4) Realize I2C (Inter-Integrated Circuit) function, and cooperate to accomplish self-system management and data transmission.

4) Fan control unit

The fan control board includes the fan and its control single board, and the main FC has three main functions: fan temperature measurement (temperature sensing function), fan speed measurement and fan speed control.

Temperature sensing can measure the internal environment temperature of the fan, and it gives information to the main control board SCTE to process through the related communication channel; The speed measurement can connect the turning data of the three fans, and it gives information to the supervision board SCTE to process through I2C interface; The fan speed control can adjust the fan speed according to the system environment so as to realize the best power dissipation and noise control.

5) Power supply unit

The power supply board has two types, one is the direct-current power unit PSA, and the other is the alternating-current power unit PSC.

(1) The input voltage of PSA is −36V~−60V, the rated power is 420 W, it supports hot plug and double power backups, which can switch the voltage from −48V to −12V and provide power to all the boards of EMB5116 TD-LTE.

(2) The input voltage range of PSC is 154 VAC~286 V AC, the rated power is 420 W, it supports hot plug, which can switch the voltage from 220 V AC to 12V and provide power to all the boards of EMB5116 TD-LTE.

6) Extend transmission processing unit

Extend transmission processing board E type (ETPE) can link S1 to realize IEEE1588V2 clock synchronization.

ETPE has the following main functions:

(1) Link 1 FE electricity to realize the information exchange function between S1/X2 and IEEE1588V2.

(2) Link 1 FE light to realize the information exchange function between S1/X2 and IEEE1588V2.

(3) Use 2 GE interfaces to connect to the SCTE sub-system to transmit the work data and control information.

(4) Output 1 PP1S and TOD information to synchronize the system.

5. The appearance of the radio frequency unit of the base station

The base station radio frequency unit includes double path RRU and 8-path RRU.

1) The appearance of the double path RRU

The double path RRU includes TDRU331FAE, TDRU332FA, TDRU341FAE, TDRU342D, and TDRU342E.

The size of TDRU342E is 420 mm × 300 mm × 120 mm (height × width × depth), the net weight of the case is 12 kg, and the equipment capacity is 15 L. Figure 2-1-5 shows its appearance.

Figure 2-1-5　TDRU342E

appearance

2) The appearance of the 8-path RRU

The 8-path RRU includes TDRU338FA, TDRU338D, and TDRU348FA.

The size of TDRU338D is 439 mm×356 mm×140 mm (height × width × depth), the net weight of the case is 23 kg, and the equipment capacity is 23L. Figure 2-1-6 shows its appearance.

Figure 2-1-6　　TDRU338D appearance

6. Service and function

EMB5116 TD-LTE can provide these services: background/conversation, fluid type, PS based dialogue (Voice over Internet Protocol, VoIP), navigation, E-MBMS (Enhanced Multimedia Broadcast Multicast Service), etc.

The functions of EMB5116 TD-LTE are as follows

1) Data service

The most distinct characteristic, which can separate TD-LTE from the 3G mobile communication system, is that TD-LTE can support high-speed data transmission. CS (Circuit Switched) domain service is cancelled, and voice can be transmitted through IP, which is VoIP.

2) E-MBMS

E-MBMS can support not only the broadcast of low-speed text messages, but also high-speed multimedia broadcast and area broadcast, and its realization is based on the TD-LTE group network. By increasing some new function utilities, for example, broadcast multicast service center, the E-MBMS function is added to the existing group domain function utilities, such as SGSN (Serving GPRS Support Node), GGSN (Gateway GPRS Support Node), eNode B and UE, etc. A new logical shared channel is defined to realize the air interface resource sharing.

3) The hierarchical structure of the high reusable software

On the software level, EMB5116 TD-LTE uses the new version of OSP (Open Systems Profile), which shields the hardware difference of all the protocols to make the software connection interface the same. This not only is convenient for software transplant and later updating, but also good for the extension of the later product series.

4) Compatibility between TD-LTE and TD-SCDMA

The design of EMB5116 TD-LTE supports TD-LTE and TD-SCDMA shared hardware platform, and the hardware supports TD-LTE and TD-SCDMA double modes. The transmission platform provides larger bandwidth, the topology structure has reasonable functions, and the connection and control parts are same. Under the TD-SCDMA mode, the baseband processing part can satisfy different setting requirements by flexible management according to TD-LTE needs.

5) Function of gathering and distribution of the baseband data of single antenna and multi-antenna

The eNode B subsystem supports the baseband resource gathering and distribution of single carrier antenna size, the flexible mapping of the baseband data between BPOX and RRU, and it makes a good use of optical fiber bandwidth and BPOX board resource to reduce the system cost. It also supports the antenna data combination to enhance the network coverage and to reduce the network construction cost.

6) Measurement

The base station needs to provide the exact physical level measurement data for LMT when TD-LTE network system deals with some key issues (for example, switching or function controlling, etc.) and optimizes the network. EMB5116 TD-LTE supports the following measurement types:

(1) receive signal code power;

(2) receive interference power;

(3) thermal noise power;

(4) noise related array;

(5) signal noise ratio (SNR);

(6) uplink synchronous location;

(7) downlink pilot frequency transmission power;

(8) CQI (uplink transmission format indication);

(9) uplink synchronous location.

7) Intelligent antenna beam forming technology

The key technology of intelligent antenna is the self-adaptation antenna beam forming, which can use the advantage of adaptive technology in beam gathering and direction control by the antenna array so as to produce several independent beams. Therefore, the beam direction can be conveniently changed to track the signal change. During receiving process, the self-adaptive weighting can adjust each element input, which is combined with other signals. It can demodulate the estimated signal out of the receiving signals and greatly reduce the interference signal. It can change the interference direction into 0 so as to reduce or even eliminate the interference signal. It can obtain the UE signal directional diagram from the receiving signals, and adjust the output intensity and phase position of each radiation array element through adaptive adjustment. Therefore, their output can be intensified to produce the purpose-oriented UE forming beam.

The intelligent antenna has great advantages in reducing interference, enlarging the area radius, lowering system cost and improving the system capacity.

8) Periodic calibration

The antenna calibration should be done periodically to realize the intelligent antenna functions, and it can compensate the deviation of the phase position and intensity among antennas.

Another function of periodic calibration is that it can detect the physical damage and wrong data transmission of the related radio frequency path.

9) Power control

In order to reduce the multi-address interference, effectively use the wireless resource of the air interface, reduce the near-far effects or the multipath fading, and guarantee all users' QoS (Quality of Service), it is necessary to strictly adjust and control UE and eNode B transmission power in TD-LTE system. EMB5116 TD-LTE can realize the power control by measuring the signal-to-noise ratio.

10) Synchronous control

The uplink synchronization is one of the most essential technologies in the TD-LTE system, and the uplink synchronization can greatly influence the whole system function. It can realize the effective use of the high frequency spectrum of the TD-LTE system.

7. A typical example of the indoor distribution and setting

When it's set as the indoor distribution signal resource, it can be jointly used with the 2-path TDRU342E. One BPOG can support three 20M2-antenna areas, as shown in Figure 2-1-7.

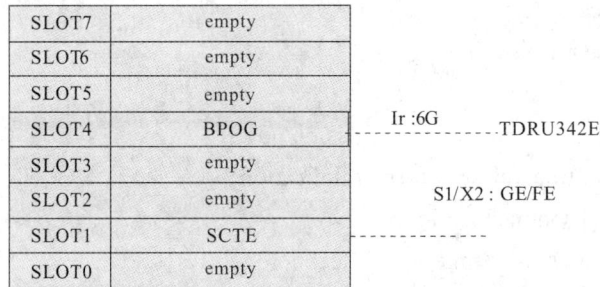

SLOT7	empty
SLOT6	empty
SLOT5	empty
SLOT4	BPOG
SLOT3	empty
SLOT2	empty
SLOT1	SCTE
SLOT0	empty

Ir :6G ---------------- TDRU342E

S1/X2 : GE/FE

Figure 2-1-7   Indoor distribution and setting

8. A typical example of outdoor macro-cellular setting

When it's set as the outdoor coverage, it can be jointly used with 2-path RRU. One BPOG can support three 20m2-antenna areas, and the setting is $1 \times$ BBU(1BPOG)+ $3 \times$ TDRU342D, as shown in Figure 2-1-8.

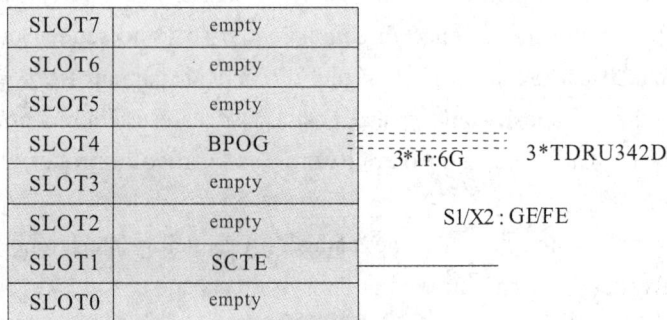

SLOT7	empty
SLOT6	empty
SLOT5	empty
SLOT4	BPOG
SLOT3	empty
SLOT2	empty
SLOT1	SCTE
SLOT0	empty

3*Ir:6G   3*TDRU342D

S1/X2 : GE/FE

Figure 2-1-8   Outdoor macro 2-path setting

One BPOG can support three 20m 8-antenna areas, and the setting is $1 \times$ BBU(1BPOG) + $3 \times$ TDRU338D, as shown in Figure 2-1-9.

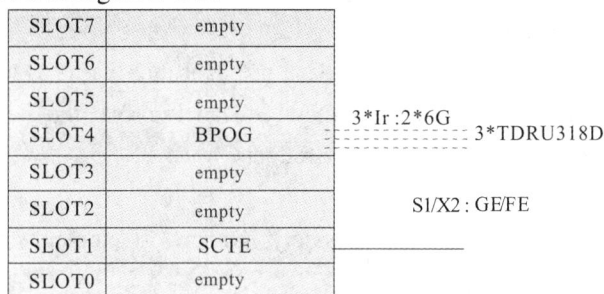

SLOT7	empty
SLOT6	empty
SLOT5	empty
SLOT4	BPOG
SLOT3	empty
SLOT2	empty
SLOT1	SCTE
SLOT0	empty

3*Ir :2*6G ------- 3*TDRU318D

S1/X2 : GE/FE

Figure 2-1-9   Outdoor macro station 8-path setting

Work Task 2: Key Network Products

For the users, LTE core network provides wide data bandwidth, low work time delay, fast connection speed, and safe IP transmission service. It supports various kinds of connection technologies, and users can successively access the network service when moving from one network into another. It also supports the emergency call and routing optimization for roaming users. It provides the flexible control strategy and billing method for service providers, and the product operation and maintenance is much easier.

1. Understanding the technical characteristics of the core network products

EPC core network is composed of different network elements, such as MME, S-GW, P-GW (PDN Gateway), PCRF (Policy and Charging Rules Function), CG (Charging Gateway), HSS (Home Subscriber Server), etc.

1) Large capacity and high integration

The product hardware platform is based on the Datang Mobile hardware and software platforms. The signal order/control interface and data interface are separated, and the several general processors with high performance can process the signal order/control interface. It uses the multicore processor technology to have the powerful ability to process the control and work interface, which can minimize the equipment cost greatly.

2) Telecommunication level reliability

The product work in the mode in which the key points consist of the primary points and the stand-by points, and the load is shared. The equipment has the high reliability by using the related technologies like failure detection,etc. location and isolation,etc. The software uses technologies like timing detection, incident report, software supervision, saving protection and resource check to improve the equipment reliability. The switching point and central control point use the standby working mode, and the connection point uses load-sharing mode.

3) Smooth expansion and flexible configuration

The product has the modular design, and uses the distributed processing and the different level exchange method. It can extend the system capacity by accumulating different modules, which can greatly save the investment. And the capacity expansion doesn't influence the existing service.

4) Compatible interfaces

The product has various types of standard protocol interfaces, which can help the product be connected with other providers' equipment. It can improve the flexibility for the service operators in choosing network construction products.

5) Safety

The safety can be divided into system safety and network element safety.

(1) The system level safety includes:

① Support the isolation function of different network safety domains.

② Support VRF isolation function (VPN Route and Forwarding Instance).

(2) The main characteristics of network element safety include the following:

① Support NAS message safety.

② Support AS safety.

③ Support verification safety.

④ Support user ID protection.

⑤ Support permanent user ID and verify vector transmission.

⑥ Adhere to key-level structure, distribution, calculation requirement.

⑦ Adhere to key caption and usage requirement.

⑧ Adhere to safe context construction requirement.

⑨ Adhere to key processing requirement for user attachment/non-attachment switching.

⑩ Adhere to key processing requirement for Idle state/connection state switching.

⑪ Adhere to mobile key management requirement.

⑫ Adhere to key update requirement.

⑬ Adhere to safe interoperation requirement between E-UTRAN (Evolved UMTS Terrestrial Radio Access Network) and UTRAN/GERAN (GSM EDGE Radio Access Network).

6) Convenient and practical operation and maintenance

(1) Operation and maintenance use the distribution design, which is deployed in the equipment operation units.

(2) Operation and maintenance use public service and adopt the partial function division mode.

(3) The reliable and safe network connection, building and processing mechanism can ensure that the equipment cannot be easily invaded.

(4) With the rich system event function and the graphical equipment supervision function, it is more convenient for the product to supervise the equipment running state.

(5) The reliable warning detection (recovery), report and transmission mechanism can ensure the network element equipment reports and transmits the warning through the sequential, no-error, and data flow control mechanism.

(6) The rich failure detection and resolution method can locate the possible source of

warnings step by step so as to rapidly find out the failure reason.

(7) The high efficient software automatic loading and online update function can greatly shorten the equipment startup delay.

(8) The simple and efficient interface for management and connection makes it convenient for the software to be updated.

(9) The flexible and useful performance measurement function, the performance statistics can be automatically adaptive to the setting change when the system setting changes.

(10) The message and signal order tracking function can locate the problem and optimize the network in practice.

7) Standard hardware platform

The product hardware platform is based on the open ATCA (Advanced Telecom Computing Architecture) structure, and it has the following features:

(1) The highly integrated board and the effective heat radiation system support a higher computing ability.

(2) Various kinds of high efficient switching and interconnected technologies can bring large data bandwidth.

(3) Realize the separation between control and service data.

(4) Satisfy the high reliability of five nines required by the operating telecommunication network.

(5) Modularization and expansion are possible for the direct update and convenient for service expansion.

(6) Have higher openness and standardization.

(7) Have good management performance and interoperability.

2. Understanding MME

The related functions of MME are as follows:

1) Access authentication function

LTE/SAE (System Architecture Evolution) network safety architecture is almost the same as that of UMTS (Universal Mobile Telecommunications System), as shown in Figure 2-2-1.

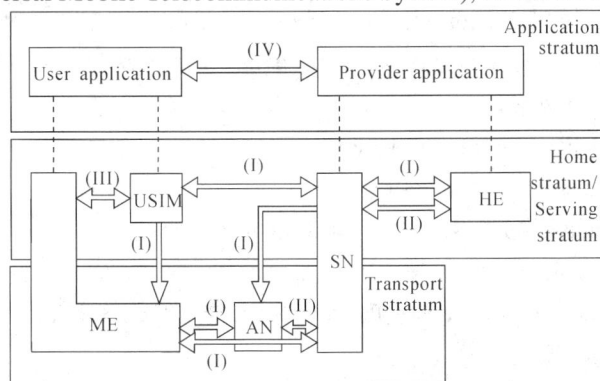

Figure 2-2-1　Safety architecture

LTE/SAE network safety can be divided into 5 domains:

(1) network connection safety (I);

(2) network domain safety (II);

(3) user domain safety (III);

(4) application domain safety (IV);

(5) visual ability and configurability of safe service.

The MME in EPC(Engineering Procurement Construction) network should have the following functions according to the above architecture:

(1) authentication function;

(2) GUTI (Globally Unique Temporary UE Identity) distribution function;

(3) user equipment identification function;

(4) signal order and data encryption function;

(5) NAS signal order encryption and consistency protection function.

2) Mobile management function

(1) MME is responsible for UE to be connected to the EPC system, which manages the mobility of the mobile phone in idle mode and tracks its location. For example, it can maintain the information about TA and MME where UE is located.

(2) The mobility management function can be realized through adherence, separation and location update, etc. These process can ensure that the UE location in the related network entity is timely updated when UE is moving, for example, the updating of the current MME information in HSS.

(3) Location management: it includes MME's distribution to TAI list according to static setting, the tracking area updates among MME: the tracking area updates among different eNode B in a MME, the tracking area updates the same eNode B in the same MME, the periodic tracking area updates, and the tracking area updates caused by the load redistribution among different MMEs.

(4) MME can impose restrictions on user mobility based on the user signed contract.

(5) MME supports multi PDN (Packet Data Network) connection: the PDN connection/release launched by UE and the PDN connection launched by MME.

(6) Support service request, including the request launched by UE and network.

(7) Support Purge process.

3) Conversation management function

MME is responsible for EPC supported construction, correction and release, and for the construction and release of E-UTRAN connection to the network.

(1) Support the following load establishment methods:

① Establish the default load of the default APN (Access Point Name) in the adherence process launched by UE.

② Correct the load resource launched by UE, and establish the special load.

③ Activate the EPC load launched by network, and establish the special load.

④ Establish the default load in the PDN connection launched by UE.

(2) Support the following methods to correct the load:

① Correct the load resource launched by UE and the special load.

② Correct the load launched by P-GW; the launched QoS has changed, and the QoS parameters and APN-AMBR need to be corrected.

③ Correct the load launched by P-GW; the launched QoS has no change, and only TFT (Traffic Flow Template) needs to be corrected.

④ Trigger the load correction launched by MME due to the contract data change in HSS.

(3) Support the following methods to delete the load.

① Correct the load resource launched by UE for special load release.

② Release the load launched by P-GW for special load release.

③ Release the load launched by MME for special load release.

All the connection load of one PDN can be released when UE or MME launches PDN to connect the process.

4) Switching control function

The MME supported switching is as follows.

(1) Switching in the E-UTRAN system: it includes EUTRAN connection to the interior part, the switching among eNode B, with MME unchanged, both changed and unchanged serving GW; and also the EUTRAN connection to the interior part, the switching among eNode B, with MME changed both changed, and unchanged serving GW.

(2) The interoperation between E-UTRAN and UTRAN or GERAN is for the core network difference in the group domain used by 2G/3G, and it has two situations. One is Rel-8 SGSN as the newest protocol which is also called S4 SGSN core network, and the other is the SGSN before Pre-Rel-8 as the newest protocol which is also called Gn/Gp SGSN. Thus, there are two network architectures for the interoperation related to LTE and 2G/3G. The non-roaming architecture is shown in Figure 2-2-2, and the roaming architecture is shown in Figure 2-2-3.

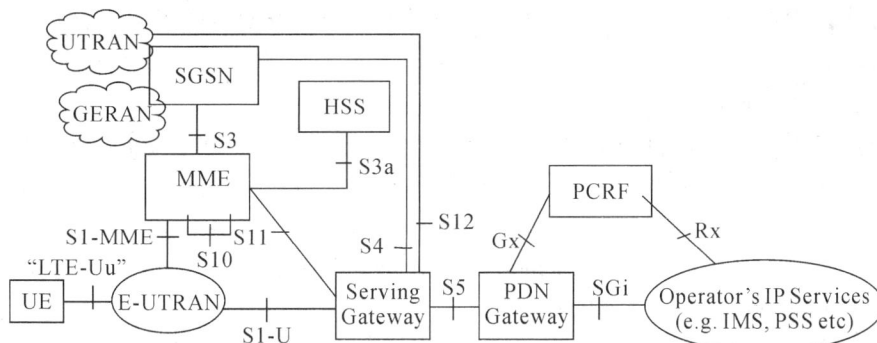

Figure 2-2-2   E-UTRAN and S4 SGSN intercommunication non-roaming architecture

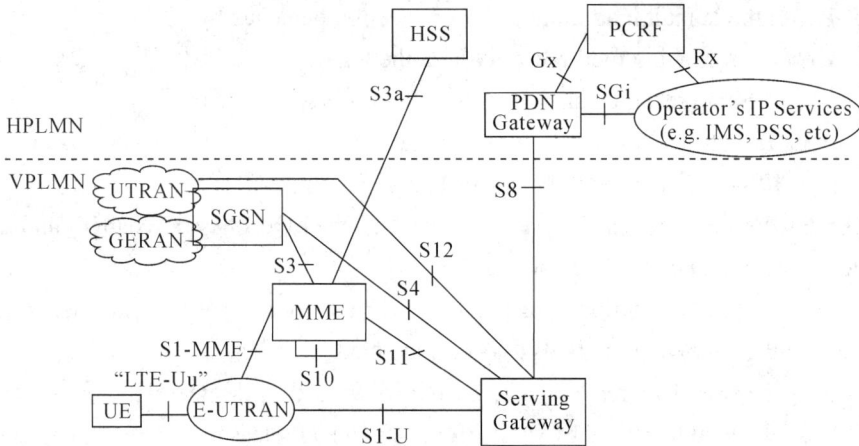

Figure 2-2-3　E-UTRAN and S4 SGSN intercommunication roaming architecture

5) MME Pool management function

In order to provide better service to operators, 3GPP has put forward the MME Pool definition. One MME Pool can serve several MMEs, each MME can save all MME IDs in the same MME Pool, and each eNode B is connected to all the MMEs in a Pool. MME Pool management has the following functions:

(1) MME supports to send its proportional message to eNode Bs in S1 setup process.

(2) MME supports to send S1 connection release message when the load transfer among MMEs happens in MME Pool.

(3) MME supports overload resolution.

6) Clock synchronization function

MME supports to set the remote server as the local time server, and its terminal works in the client mode.

The network time protocol NTP (Network Time Protocol) is the ICP/IP protocol used to give the exact time in the whole IP network, and its transmission is based on UDP. The RFC1305 regulation has given the complex algorithm used by NTP to ensure the accuracy of clock synchronization.

7) GTP-C function

Both the S11 reference point between MME and S-GW and the protocol reference point S3 between MME and SGSN use the GTP tunnel protocol to transmit data. The GTP-C of S11 interface can be used to transmit the MME control signal order for user interface, and GTP-C of S3 interface can be used to support the mobility between 2G/3G system and LTE system. Figure 2-2-4 shows the protocol stack organization.

The specific functions are as follows.

(1) Achieve GTP-C signal order message encoding and decoding.

(2) Create GTP tunnel.

(3) Delete GTP tunnel.

(4) Update GTP tunnel.

(5) Achieve GTP-C message transmission.

8) SGs interface short message transmission function

SGs is the interface between MME and MSC, which can be used to negotiate 2G/3G conversation and transmit short messages. SGs interface protocol stack organization is shown in Figure 2-2-5.

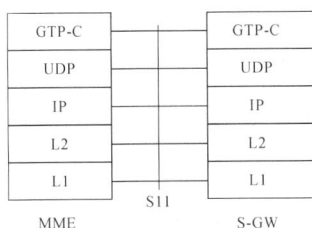

Figure 2-2-4　MME supported
S11/S3 interface protocol stack

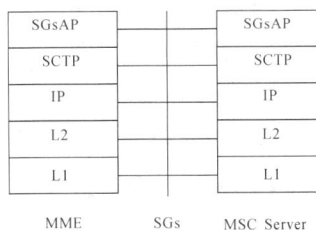

Figure 2-2-5　SGs interface

Interface protocol instruction is as follows:

(1) SGsAP (SGs Application Protocol): this protocol can connect MME and MSC, maintain signal order communication during 2G/3G conversation and transmit short message.

(2) SCTP (Stream Control Transmission Protocol): a signal order transmission protocol.

9) Technical specification

MME (TLE3200) capacity specification is shown in Table 2-2-1.

Table 2-2-1　TLE3200 capacity specification

Project	specification
Supported user number	6 000 000
Supported EPC load number	12 000 000
Equal BHCA	39 000 K
Time delay of call establishment processing	<40 ms
Time delay of call release processing	<40 ms
Maximum number of linked eNode B	5000
Maximum number of linked SGW	128
Maximum number of linked MME	128

MME (TLE3200) reliability specification is shown in Table 2-2-2.

Table 2-2-2　TLE3200 reliability specification

Project	Specification
MTBF (Mean Time Between Failure)	≥380 000 hours
MTTR (Mean Time To Restoration)	≤30 minutes
Usability	≥99.999%
Mean break time in a year	≤3 minutes

The operation and maintenance specification of MME (TLE3200) is shown in Table 2-2-3.

Table 2-2-3   TLE3200 operation and maintenance specification

Project	Specification
Warning saving time under disconnection with OMC	≥7 days
Specification data saving time under disconnection with OMC	≥7 days

3. The specific functions of S-GW:

1) Load management function

(1) Load establishment goes as follows.

① Make load resource correction launched by UE.

② Establish the default load of default APN in the Attach process launched by UE.

③ Make EPC load activation launched by network.

④ Establish default load upon EU's request to which several PDN connetions be established.

(2) Load correction goes as follows:

① Make load resource correction launched by UE.

② The load QoS changes when P-GW launches load correction, which includes the correction of QoS parameters and APN-AMBR (Aggregate Maximum Bit Rate) .

③ The load QoS does not change when P-GW launches load correction, and in this case only TET is corrected.

④ The contract data change in HSS can trigger MME to launch load correction, which includes the correction of APN-AMBR and UE-AMBR.

(3) Load release goes as follows.

① Process load resource correction launched by UE.

② Process load release launched by P-GW.

③ Process load release launched by MME.

④ Delete all the load of the PDN connection when UE requests that several PDNs be connected.

(4) Network triggers the service request.

When S-GW receives the downlink data and S1-U interface load information is released, S-GW gives notice to MME to launch a call.

2) Router choice and data transmission functions

(1) S-GW should have the function to transfer the received data (GTP-U PDU) from the previous point to the middle or the lower point of the router.

(2) After the switching among eNode Bs or systems, S-GW user interface should send "end marker" data packet to source eNode B, source SGSN or source RNC (Radio Network Controller) to realize the rearranging function of eNode B.

3) QoS control function

(1) S-GW supports the main QoS parameters loaded by EPC, which includes QCI (Quality of service Class Identifier), ARP (Address Resolution Protocol), GBR (Guaranteed Bit Rate),

MBR (Maximum Bit Rate) and APN-AMBR.

(2) S-GW supports the load correction process based on QoS and launched by terminal and network.

(3) S-GW supports to realize the MBR bandwidth management function of the bearer level for GBR load.

(4) S-GW supports the DSCP marking function of uplink and downlink data based on the bearer level.

(5) S-GW supports the reflection relation between QCI settings and QoS parameters.

4) GTP function

The GTP tunnel function of S-GW is mainly used for the S11 interface between S-GW and MME. The S5/S8 interface has the data transmission function on the user interface when the GTP tunnel protocol is used between S-GW and P-GW, and the control interface is used for tunnel management.

5) Charging function

(1) offline charging;

(2) load-based charging;

(3) setting of charging rules.

6) Technical specification

S-GW (TLE3300) capacity specification is shown in Table 2-2-4.

Table 2-2-4   TLE3300 capacity specification

Project	Specification
Supported user number	1 000 000
Supported EPC load number	2 000 000
Equivalent data throughput capacity (double direction)	100 GB/s
Load establishment processing time delay	< 40 ms
Load release processing time delay	< 40 ms
Present service time delay	< 2 ms

TLE3300 reliability specification is shown in Table 2-2-5.

Table 2-2-5   TLE3300 reliability specification

Project	Specification
MTBF	≥ 380 000 hours
MTTR	≤ 30 minutes
Usage	≥ 99.999%
Mean break time in a year	≤ 3 minutes

TLE3300 operation and maintenance specification is shown in Table 2-2-6.

Table 2-2-6　TLE3300 operation and maintenance specification

Project	Specification
Warning saving time under disconnection with OMC	$\geqslant 7$ days
Specification data saving time under disconnection with OMC	$\leqslant 7$ days

4. Understanding P-GW

1) UE IP address management function

(1) P-GW can be distributed locally.

(2) P-GW supports to set different address pools according to different APN.

(3) P-GW should remove the enclosed head to the outside data network if GTP-U PDU goes to the outside data network.

2) Router choice and data transmission function

(1) P-GW should have the function to enclose the GTP head and UDP/IP head of PDU from the outside data network. By taking the related head address as the indication information, it will use one point-to-point double direction tunnel in the EPC network to transmit the enclosed data to the terminal.

(2) P-GW should remove the enclosed head to the outside data network if GTP-U PDU goes to the outside data network.

3) QoS control function

(1) P-GW supports EPC to load the main QoS parameters which include QCI, ARP, GBR, MBR, and APN-AMBR.

(2) The network can distribute the load level QoS parameters of the default and primary load according to contract data.

(3) P-GW supports to set PCC (Policy and charging control) locally.

(4) P-GW supports the special creation and the correction load launched by UE and network, only the EPC can decide whether it's established or not, and QoS parameters of the load level are distributed by EPC.

(5) P-GW supports to launch the load correction process based on the QoS update.

(6) P-GW supports the MBR bandwidth management function of the bearer level for GBR load.

(7) P-GW supports the APN-AMBR bandwidth management function (Policing) for the uplink and downlink data flow of non-GBR load.

(8) P-GW supports the DSCP marking function of the bearer level, and the operator can set the DSCP value of QCI.

(9) P-GW supports to set the mapping relation between QCI and QQS.

4) Safety function

(1) Support the different safety domain isolation function in the network.

(2) Support VRF isolation function.

5) GTP function

S5/S8 between P-GW and S-GW is based on the GTP tunnel protocol, the user interface supports the data transmission function, and the control interface is used for the tunnel management function.

6) Charging function

(1) offline charging;

(2) service data flow charging;

(3) setting of charging rules.

7) Technical specification

P-GW (TLE3310) capacity specification is shown in Table 2-2-7.

Table 2-2-7   TLE3310 capacity specification

Project	Specification
Supported user number	1 000 000
Supported EPC load number	2 000 000
Equivalent data throughput capacity (double direction)	100 GB/s
Load establishment processing time delay	< 40 ms
Load release processing time delay	< 40 ms
Present service time delay	< 2 ms

TLE3310 reliability specification is shown in Table 2-2-8.

Table 2-2-8   TLE3310 reliability specification

Project	Specification
MTBF	$\geqslant$ 380 000 hours
MTTR	$\leqslant$ 30 minutes
Usage	$\geqslant$ 99.999%
Mean break time in a year	$\leqslant$ 3 minutes

TLE 3310 operation and maintenance specification is shown in Table 2-2-9.

Table 2-2-9   TLE3310 operation and maintenance specification

Project	Specification
Warning time saved under disconnection with OMC	$\geqslant$ 7 days
Specification data time saved under disconnection with OMC	$\geqslant$ 7 days

5. Understanding PCRF

The policy and charging control (PCC) technology means that when the user service data flow is transmitted through PS domain, the network element PCRF(Policy and Charging Rules

Function) can carry out the dynamic QoS and charging policy control timely for the user-level service data flow, which is based on the data flow characteristic, operator policy and contract user features. The process can help the operator effectively control and manage the user service.

PPC works on the service data flow, PCRF can use PCC rules to provide setting solutions, charging control, and event report function for the the service data flow.

Its main function can be divided into the following sub-functions:

(1) binding mechanism characteristics;

(2) event triggering characteristics;

(3) policy control characteristics;

(4) service first and conflict processing characteristics;

(5) charging rule characteristics;

(6) management of PCRF contract data characteristics;

(7) PPC rule interaction characteristics among roaming PCRFs.

The above characteristics can be realized through IP-CAN dialogue establishment, correction and termination.

1) Special function of binding mechanism

The binding mechanism refers to the process for binding the service data flow and its IP-CAN load. The service data flow is defined by the service data flow template (SDF template) of the PCC rule. Thus, the binding mechanism is to bind the AF dialogue message with the transmission service data flow IP-CAN load.

The binding mechanism has the following three processes.

(1) The dialogue binding by PCRF: bind the IP-CAN dialogue with the AF dialogue message (that is the applicable dialogue message provided by P-CSCF) or the predefined dialogue message, and the used PCC rule. Thus, the PCC rule of IP-CAN can be produced.

(2) PCC Rule authentication and QoS rule establishment by PCRF: choose the QoS parameters (QCI, GBR, MBR, ARP, etc.) for the PCC rule of IP-CAN dialogue. PCRF can reevaluate and authorize the PCC rule when some conditions change.

(3) Load binding: bind the designated PCC rule and QoS rule with one of the IP-CAN load in the IP-CAN dialogue. In the 3GPP-EPS system, the load QoS and the load launching control is on the NW side. During the resource request process launched by network and different UE, the load binding for these two modes are carried out in PCEF.

2) Special function of event trigger

RCRF defines the condition for PCEF (Policy and Charging Enforcement Function) to interact with PCRF again after the IP-CAN (Connectivity Access Network) dialogue establishment. PCEF will receive the condition information from PCRF.

PCRF uses PCC customization procedures to provide event trigger to PCEF. The event trigger is related to all PCC rules of one IP-CAN, and the event trigger can decide the time when PCEF notifies that the IP-CAN load has been corrected. PCRF can ask PCEF to react to all the event triggers listed in Table 2-2-10.

Table 2-2-10   Event trigger list

Event trigger	Description	Entity needed to be reported	Report condition of setting entity
PLMN (Public Land Mobile Network) change	UE moves into another operator domain (note 2)	PCEF	PCRF
QoS change	QoS of IP-CAN load changes (note 3)	PCEF, BBERF	PCRF
QoS change surpassing authentification	QoS of IP-CAN has already changed and surpassed the authorized QoS (note 3)	PCEF	PCRF
Contract APN-AMBR change	Contract APN-AMBR change	PCEF, BBERF	PCRF
QoS change of EPS contract	QoS change of default EPS load	PCEF, BBERF	PCRF
Service reflection message change	Service reflection message of IP-CAN load has already changed (note 3)	PCEF	Unconditional setting
Resource correction request	BBERF or PCEF receives a resource correction request trigger (note 6)	PCEF, BBERF	Unconditional setting
IP-CAN type change (note 1)	Access type of IP-CAN load has already changed	PCEF	PCRF
RAT (Radio Access Technology) type change	Change of wireless interface feature	PCEF, BBERF	PCRF
Lose/recovery of the transmitted resource	IP-CAN transmitted resource has no use or can be used again	PCEF, BBERF	PCRF
Location change (service area) in EPS (Evolved Packet System): (1) ECGI change; (2) CGI/SAI change	UE service area change	PCEF, BBERF	PCRF
Location change (service area) in EPS (Evolved Packet System): (1) Routing area change; (2) Tracking change; (note 4)	UE service area change	PCEF, BBERF	PCRF
Location change (service area) in EPS (Evolved Packet System): S-GW change (note 5)	Core network point change of UE service	PCEF, BBERF	PCRF
Credit used out	No effective credit point	PCEF	PCRF
Executed PCC rule request	PCEF executes PCC rule request according to PCRF indication	PCEF	PCRF

Continued Table

Event trigger	Description	Entity needed to be reported	Report condition of setting entity
UE IP address change	Distribute one UE IP address or release one UE IP address	PCEF	Always setting
Flow report	PCEF reports data flow according to PCRF indication	PCEF	PCRF

Note 1: this list doesn't contain all conditions. The exact event of each IP-CAN should be appointed later.

Note 2: IP-CAN change can lead to possible PLMN change.

Note 3: load binding mechanism is effective only when it is executed by PCRF.

Note 4: service domain change may lead to the service area change and core service network change.

Note 5: change of core service network point may lead to service area change and possible service domain change.

Note 6: for the independent resource request supported and loaded by IP-CAN network, the event trigger is effective. EPS network also belongs to this kind of network which supports the independent resource request of load.

If the location event trigger is set, the related IP-CAN process reports and triggers the location change indication. If credit authentication and event trigger need to report different level location change, the reported location level should be changed into the highest level required. If the reported level can give more detailed information than the request of PCRF, the update process of PCC rule or QoS rule should not be triggered.

3) Special function of strategy control

Strategy control specification includes three parts: gating control, event report, and the establishment/correction/termination of IP-Can load. PCRF executes gate control decision, event report resolution and QoS control strategy.

(1) Gating function is based on each service data flow. PCRF can make the gating decision according to the AF reported dialogue, and allow service data flow to pass or stop by giving notice to the designated point. Dialogue event includes dialogue termination and correction. The execution location of gating function is on PCEF.

(2) Event report: it refers to PCRF notice and AF application event notice, or GW, GW(BBERF) resource related event report, and it can trigger QoS control process to update the user interface operation. The event report mechanism is related to the event trigger notice mechanism;

(3) QoS control: PCRF dynamically customizes QoS authentication message according to PCEF and BBERF (Bearing Binding and Event Report Function), which is based on various events.

4) Special function of service priority and conflict resolution

When one user successively requests to activate several service and the service bandwidth accumulated by several PCC rules, which is larger than the guaranteed bandwidth contracted by the user, the PCRF may use the pre-emption priority of a service to decide whether any one or several of the lower priority PCC rules should be activated or not under the QoS mode. It will activate the higher priority PCC rules to ensure the contract guaranteed bandwidth is not surpassed. If the previous several services are not allowed to be occupied, PCRF will deny the last PCC rule for its activation service request.

5) Special function of charging rule

In order to realize the charging function based on service data flow, PCRF appoints the charging rule based on the service data flow, contains the charging rule for the message related to strategy control, and gives it to PCEF. PCEF will evaluate the charge based on the charging rule.

The charging rule includes service data flow template, service data flow template priority, charging key (the related charging group), service identifier, charging method, measuring method, AF record information (the usage terminal reports the related charging key and service identifier), and the report information of service level. It can detect and measure the data packet of service data flow.

The measuring methods for the above charging rule are as follows:

(1) charging measurement method based on data flow;

(2) charging measurement method based on time;

(3) charging measurement method based on the combination of data flow and time;

(4) charging measurement method based on event;

(5) no designated charging rule.

Charging measurement method based on event doesn't belong to the PCC rule designated by PCRF, and it is the predefined PCC rule in PCEF.

PCRF doesn't provide the above charging measurement methods to PCEF which can use its predefined measurement method.

6) Special management function of PCRF contract data

When the IP-CAN dialogue is established, PCRF will request SPR to provide the user with contract information, and PCRF doesn't need to save the contract information used for PCC decision until the IP-CAN dialogue stops. PCRF needs to book the notice requirement of the contract information change from SPR. If PCRF receives the contract data change notice, then it needs to update the PCC rule based on the updated contract data. When the related contract information is deleted, PCRF should send a request message to SPR to cancel the notice.

7) Technical specification

PCRF (TLE3400) capacity specification is shown in Table 2-2-11.

Table 2-2-11　TLE3400 capacity specification

Project	Specification
Supported user number	Maximum to 6 000 000
Time delay for PCC rule processing	< 20 ms

TLE3400 reliability specification is shown in Table 2-2-12.

Table 2-2-12　TLE3400 reliability specification

Project	Specification
MTBF	≥ 380 000 hours
MTTR	≤ 30 minutes
Usage	≥ 99.999%
Mean break time in a year	≤ 3 minutes

TLE3400 operation and maintenance specification is shown in Table 2-2-13.

Table 2-2-13　TLE3400 operation and maintenance specification

Project	Specification
Warning time saved under disconnection with OMC	≥ 7 days
Specification data time saved under disconnection with OMC	≥ 7 days

6. Understanding CG

The charging gateway is one of the important network elements in TD-LTE EPC core network, which can process the dialogue after receiving the original dialogue record from S-GW and P-GW, and create the dialogue file finally. It can use FTP (File Transfer Protocol) mechanism to transmit the final dialogue file to BOSS (Business & Operation Support System) to process. The related function includes dialogue collection (receive the original dialogue, verify the dialogue, save the dialogue and detect interface link, etc.), dialogue process (carry out the dialogue rearrangement and dialogue combination, etc. for the original dialogue received), and dialogue file management (save the final dialogue as a file, provide the automatic dialogue copy and the automatic dialogue deletion, etc).

1) Dialogue collection function

The dialogue collection function is mainly composed of dialogue collection process and dialogue saving process. The dialogue collection process is responsible for receiving the original dialogue sent by CDF (Communication Data Field) from Ga interface, verifying the dialogue and inserting the dialogue into the storage area for the dialogue storage process to read. The dialogue collection process still needs to detect the Ga interface link state and gives warning messages when the link has some faults. The dialogue saving process is responsible for reading dialogue from the storage area and inserting the dialogue into the database. Then, it sends the response message to the CDF terminal.

CG not only supports the dialogue transmitted by several CDFs, and also processes different types of dialogue. The supported dialogue types are shown as follows:

SGW-CDR refers to dialogue produced by S-GW.

PGW-CDR refers to dialogue produced by P-GW.

The original dialogue storage function is called the original dialogue, which refers to the unprocessed dialogue received from CDF. CG needs to save the original dialogue through different files, which can save the original dialogue, the wrong dialogue and the final dialogue into different files. The user can visit the dialogue list through dialogue-visiting tool.

2) Dialogue processing function

Dialogue processing function is mainly finished by a dialogue processing program, which mainly includes dialogue duplication exclusion and dialogue combination.

Dialogue duplication exclusion means to abandon the repeated original dialogue received, and only saves one copy. The repeated dialogue means that the dialogue P-GW address, S-GW address, CID, recorded type, and Record Sequence Number, are all the same.

Dialogue combination means the several partial dialogues which belong to the same load are combined into a final dialogue. For the same IP-CAN load context, it can produce several partial dialogues due to the QoS change, charging time change, data flow restriction, time restriction, charging condition change, etc. CG can combine these dialogues into one dialogue according to the related rules. The context loaded by IP-CAN one time belongs to one P-GW, and it is simple to combine the partial PGW-CDR for one P-GW. The SGW-CDR dialogue combination is much more difficult due to S-GW switching, and CG can flexibly combine the different dialogues and finish the dialogue combination function.

For SGW-CDR combination, it supports the partial dialogue combination of SGW-CDR produced by S-GW.

3) Dialogue file management function

The dialogue processing program can finish the dialogue file management function, which includes the final dialogue establishment, the automatic copy of the final dialogue and the automatic deletion of the final dialogue.

The dialogue record after the dialogue combination needs to be created as a dialogue file which can be transmitted to BOSS system through FTP protocol, and CG needs to save the final dialogue file locally for a while. The final dialogue can be produced timely and quantitatively. The time production means one final dialogue file can be produced in a regular time, and quantitative production means a final dialogue file can be produced according to its maximum size.

The automatic dialogue copy means to copy the dialogue file when it meets some requirements, which can be saved into the another logic division of the local disk. The automatic copy means the program can automatically carry out the dialogue copy operation under the related condition.

Automatic dialogue deletion means to delete the related dialogue file according to some requirements, which normally refers to the deletion of the dialogue file which finishes copying. The program can automatically delete the dialogue under the required condition.

4) Technical specification

CG (TLE5600) interface capacity specification is shown in Table 2-2-14.

Table 2-2-14 TLE5600 interface capacity specification

Project	Specification
Optic interface	4
Electric interface	4

Note: optic interface can be used for Ga/Bp interface, which uses the 1+1 main and standby method.

TLE5600 service capacity specification is shown in Table 2-2-15.

Table 2-2-15 TLE5600 service capacity specification

Project	Specification
processing capacity of Peak value dialogue	500 CDR/S
saved time Original dialogue	90 days
time Final dialogue saved	90 days
Dialogue storage capacity	800 G

TLE5600 reliability specification is shown in Table 2-2-16.

Table 2-2-16 TLE5600 reliability specification

Project	Specification
Mean time between failures (MTBF)	$\geqslant$ 100 000 hours
Mean time to repair (MTTR)	$\leqslant$ 30 minutes
Usage	$\geqslant$ 99.999%
Mean break time in a year	$\leqslant$ 3 minutes
Message loss probability	$\leqslant$ 0.0001%
System cold-starting recovery time	< 30 minutes

TLE5600 operation and maintenance specification is shown in Table 2-2-17.

Table 2-2-17 TLE5600 operation and maintenance specification

Project	Specification
Warning time saved under disconnection with WebServer	$\geqslant$ 7 days
Specification data time saved under disconnection with WebServer	$\geqslant$ 7 days

The technical specification of TLE5600 server hardware is shown in Table 2-2-18.

Table 2-2-18   Technical specification of TLE5600 server hardware

Project	Specification
Size	85.4 mm × 443.6 mm × 798.0 mm
Weight	21.1–24.9 kg
Power	675 W
Processor	Quad-core Xeon E5620 2.4 GHz 2
RAM	4 GB PC3-10600 CL9 ECC DDR3 1333 MHz RAM 16
Network card	Integrated dual-port Gigabit Ethernet, Gigabit dual-port Ethernet card × 1, single-port Gigabit optical fiber card × 4
Hardware	300 GB 10 K 6 Gbps SAS (SFF) thin hot plug hardware × 4 (RAID5+1 backup disk)
Power-supply module	675 W hot plug DC power × 2 (or: 675 W hot plug AC power × 2: support double-path 110 V-220 V AC power supply)
Others	DVD driver × 1

7. Understanding HSS

The Home Subscriber Server is an important network element in the LTE EPC, which includes the user setting information, executes the user ID verification and authentication, and provides the user with the related physical location information. It can complete the user information contract in LTE and user authentication function, etc. The specific functions are as follows:

1) EPS user accessibility management function

(1) Set accessibility management function in the user contract data: give notice to the related MME.

(2) Save the accessibility information of the application user, and gives the user accessibility notice to the related service platform.

2) Function for processing the Notify request launched by MME

It can process the Notify request launched by MME, and can carry out the related operation based on the specific information in the request, for example:

(1) Update the terminal information.

(2) Send the location deletion order to the present MME.

(3) Set the restricted access for the present domain.

(4) Send UE accessibility notice.

3) Mobility management function

TLE5500 can evaluate the user roaming state and judge whether the user roams out of the operator's network. Based on the above judgement, if the required condition is satisfied, the operator for the users can define the service type, and it doesn't send the contract parameters or the default value to MME.

TLE5500 can control the user access type based on the contract parameters.

(1) TLE5500 saves the MME address which serves the present user and the related network parameters of the MME.

(2) TLE5500 can cooperate with MME for launching the location registration/logout notice to finish the user location registration/logout state and updating the present MME address. TLE5500 supports the double registration of MME.

(3) Under the following condition, TLE5500 will actively launch logout request to MME and bring the related logout type information: switching to the new MME of the user, the forced change by network of the user registration state, the MME address, user deletion, etc.

(4) After receiving the UE request from MME, TLE5500 can set the "UE clearance" marker to UE.

4) Support function for area contract restriction

TLE5500 supports the area contract restriction service based on RSZIs. Under the RSZIs realization mode, the service department has to define RSZIs (defined as province, mobile local network or special area) according to the service needs. The user can put forward his/her request to the roaming area according to his/her real situation when signing contract with the operator, and the roaming area is marked by the area contract identifier. Each user can be assigned 10 area codes at most. After signing the contract, the operator can provide the related area roaming service to the user, which is based on the user contract area code sequence. TLE5500 gives the user contract area code in the location update information and the user contract data.

5) User data management function

It is necessary that user data should be operated and managed by TLE5500 according to actual needs.

(1) Open or cancel account;

(2) User contract data correction which includes the newly increased contract service data and the contract service data correction;

(3) Support the batch processing of the user data, which supports to complete the huge amount of similar operation on large number of users;

(4) TLE5500 can notice that MME updates the user contract data (increasing or replacing some users' data);

(5) TLE5500 can give notice to MME to delete partial saved data of the user;

(6) Check/increase/delete/correct the user identifier and contract data.

6) Authentication support function

TLE5500 contains the function of authentication center, and can provide a group or several groups of authentication parameters to MME according to its request. It supports to process the authentication service.

7) Safety management function

Safety management can provide the effective control to restrict the user access or user visit to TLE5500. It can ensure that each of the legal users is possible to log in, use the authorized software module, complete the legal operation order, and it stops the unauthorized visit. It can protect the stable operation of the network equipment, and record the verification occurrence in the system or the authorized visit, etc. Therefore, any operation has the property of non-repudiation.

8) Recovery function

The user data is available to outside equipment, such as a disk, at regular times. The user data can be manually recovered when the system fails and data is lost. It can copy the data in TLE5500 periodically.

After TLE5500 restarts, it will execute the related program to recover TLE5500, which bases on the previous copy. It can obtain the correct mobile user location and additional service information, and give notice to the current MME for the influenced users.

The system can automatically record the input operation order. After the system reloading, it can compensate and reload the related data based on the previous order record.

9) Compatibility support function

The backward compatibility supports the Diameter protocol defined by 3GPP.

10) Technical specification

Table 2-2-19 and Table 2-2-20 shows the technical specification of HSS (TLE5500).

Table 2-2-19　Technical specification for simple setting

Technical specification	Specification value
User number	10 000
Information processing capacity(per 10 records)	1 /s
Information loss ratio	$P \leqslant 10^{-7}$

Table 2-2-20　Technical specification for standard setting

Technical specification	Specification value
User number	100 000–1 000 000
Information processing capacity	100–1000/s
Information loss ratio	$P \leqslant 10^{-8}$

Work Task 3: OMC Network Management Products

1. OMC products

OMC can provide management in setting, warning, specification, safety, daily record and software, etc. for the connected equipment, as is shown in Figure 2-3-1.

Network management is responsible for maintaining the telecommunication network equipment. It provides a man-machine interface. It can configure, check, control, and diagnose telecommunication network equipment as well as check their operation authority, track the operation, collect and analyze the operation data.

OMC network management can manage and maintain various network elements, such as wireless equipment eNode B, core network equipment EPC, and IMS (IP Multimedia Subsystem). It is compatible to manage and maintain RNC and Node B which are connected to 3G network.

OMC can provide a timely supervision function to the network element equipment and the whole network, the simple and direct data setting, the software update and system extension. It can also provide a comprehensive performance statistics function, and support all kinds of users for their convenient and efficient work.

(1) Network optimization work: collect the network data and optimize the setting of the OMC system;

(2) Engineering maintenance work: supervise, maintain network equipment and set data to the OMC system, complete the starting of network element equipment, locate and clear failures, expand the network and provide centralized upgrade for network equipment.

(3) Upper level network management: provide the network element-level operation and maintenance service of log alarm, performance specification, parameters setting, etc. through OMC system.

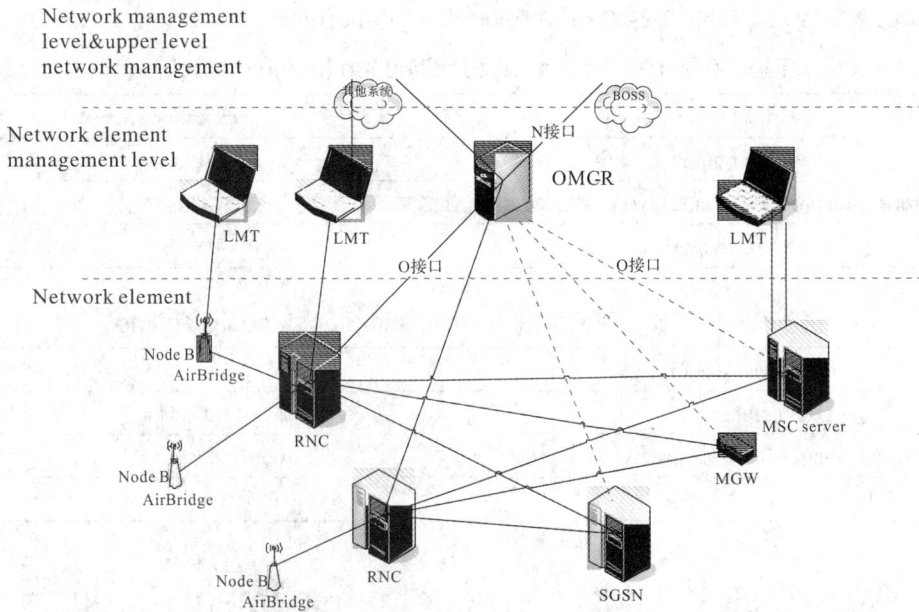

Figure 2-3-1　OMC network management system

2. Technical characteristics of OMC products

1) Centralized network management(large capacity)

OMC is mainly used to manage several equipment, and supervise the whole network operation state. Thus, it has more powerful a multi-event processing ability and a performance statistics ability compared with LMT (Local Maintenance Terminal), which can be used in the network maintenance and optimization.

Besides realizing the centralized network management function, the OMC system of

Datang Mobile also integrates part functions of LMT, which can not only satisfy the needs of the network management work, but also provide the support for related special applications.

2) Rich function and convenient operation

OMC system supports centralized management of all kinds of Datang Mobile network elements, such as eNode B and all elements in EPC and IMS. The software can be used to fulfill the related tasks through the graphic interface, for instance, warning management, setting and state management, performance statistics, security and record management, software management, topology management, system management, detection and diagnosis, and MML.

3) Fault tolerance

(1) The reported message from the equipment which cannot be identified will be abandoned and has no influence on later information processing.

(2) If there is some information in the data file which is difficult to analyze, the system will try its best to process it so that the effective data can enter the database.

(3) If the values are beyond the value range when checking the parameters of the related items, the system can display their original values and highlight them.

(4) If there is operation failure in implementing a single order during its operation process, the system can terminate the order, indicate or jump over the failure.

(5) If there are mistakes in program processing, all the information about the mistakes can be captured and timely recorded, which only influences the present event operation.

4) Maintainability

(1) It can provide log hierarchical logging function, which can record the abnormal operation of the system, key process information, and stages exchange among outside equipment.

(2) Provide one-click operation to get the running log.

(3) Support remote check on the system resource usage.

(4) Support to load and update in patch mode without restarting.

5) Recovery ability

The system supervision process can actively activate the server when it exits abnormally. For example, the external communication system (including network element, upper level network management) can be recovered if the communication stops.

6) Easy installation

Different operation systems need specialized software installation guidance, such as an installshield tool and WebStart technology, which can easily help to complete the software installation. It also has a detailed software installation guidance book. Client software can be installed or run.

The automatic version comparison can be carried out during hardware update, and it can download the updated incremental patches and version.

OMC can use the InstallShield tool to install hardware, and the process is much simpler. It can use WebStart technology for the user software installation, which can solve the client update requirements at the same time.

7) Adaptability

The OMC system can support all kinds of main softwares, operation systems and databases.

The use of J2EE technology can use InstallShied to produce the installation package for different hardware platforms. According to the hardware platform encapsulation, differentiated function can be realized, such as system supervision and FTP visit.

8) Internationalization

(1) The system supports multi-national languages, and it can automatically choose the display language and time zone according to the area setting of the system. It supports the simplified Chinese and English at least.

(2) Any display can be realized through the international encoding rules provided by the platform.

9) OEM

(1) Through the simple change of file and picture setting, it can produce the same level baseline OEM (Original Entrusted Manufacture) product.

(2) The system can acquire all the related displaying resource according to the platform definition. It encloses the different resource package of different companies during the resource acquisition process.

10) Testability

The software testing process tests the software design and execution to find out the software problem, isolate the problem and locate the problem with a certain amount of time and cost. Simply speaking, the testability of software refers to the simplification level for testing the computer program.

The features of the software testability mainly refer to set the observing point, control point, observation equipment, driving equipment, and isolation equipment. Great attention should be given to the fact that the testability design should guarantee that it doesn't influence any of the system functions, produce any additional activities or additional tests. It should use the suitable designing mode to design the software.

The following methods can be used to guarantee the testability:

(1) object-oriented design, and object-oriented language JAVA chosen for design;

(2) data display, control and isolation;

(3) modulation design;

(4) interface-oriented design;

(5) direction-oriented design (to be realized through OP (Operation Platform));

(6) changeable setting design of the service process (to be realized by introducing the work flow engine);

(7) standardization of log output.

11) Management and storing capacities

(1) Network management capacity. OMC system supports to manage the core network and the connected equipment at the same time; under the standardized setting condition, it can satisfy

all the equipment management needs for bigger or middle-sized cities.

(2) OMC management capacity. It can be improved by increasing the number of service servers and network element adapters.

(3) Data storing capacity. OMC can store specification data for three months, warning data for three months, user daily operation data for 12 months, and specification tables for 3 years, at least.

During the design of OMC setting lists, data storage space needed should be calculated according to the network size and service data flow, together with another additional 20% space.

12) Specification needs

(1) Data synchronization time. The synchronization time for a single network element is <= 1min. As a key technical parameter, it needs to be satisfied in the designing stage of configuration management.

(2) Warning receiving capacity;

(3) Guarantee the reliable processing in sequence;

(4) Have the ability to process 300 pieces per second;

(5) The time delay of displaying a warning between the time reported at a network element and the time displayed at the terminal should be no more than 5 seconds, as a key technical parameter, it needs to be satisfied in the designing stage of the warning management.

13) Safety

(1) Operation confirmation function. The user should confirm the important correction and deletion twice before the related orders are executed. For some important operation which has a great effect on the system, such as the system resetting, the result of the related operation should be displayed in the indication dialog box.

For the safety management, whether the operation needs confirmation or the password should be configured globally.

(2) Message encryption function. In order to prevent the attacks from hackers on network element equipment, the message for OMC and network element interface should be encrypted. The common algorithm can be used as the encryption algorithm, such as SSL (Secure Socket Layer) and MD5. Whether use SSL interface encryption can be configured for the internal communication of OMC. The north-direction interface can choose whether use SSL encryption according to the related regulation.

Safety channel technology is used between OMC and network element, such as the user authorized SNMPV3 protocol.

(3) Testing function for abnormal processes and interfaces. It supports to set trust lists for a process or an interface, and gives warnings if the starting process and the used interface are out of the list. It can be designed and realized as the system management function.

14) Openness

It supports the data openness needs of setting, specification, warning and log, and also supports data sharing at all levels, such as database-level and file-level. OMC can export the general data into files (like CSV, XLS or XML, etc.), and support the graphic sharing of database. Thus, it can share the data according to the needs.

3. System structure of OMC products

The typical LTE-OMC system, and its hardware includes OP platform, OP-BA, user end, server and outer interface.

1) OP platform

OP platform is designed and developed for OMC network management products and OSS integrated system products. These two products have many sub-types and complicated functions, and they have strict requirements for stability and specification. The products have many kinds of special logics and general service logics of software for their huge amount of data. If they cannot be divided effectively and developed into specialized products, there will be much more development difficulties, longer development period and lower development efficiency, etc.

The purpose for designing the OP platform is to strongly support these two kinds of systems, which can provide clear, rich, stable and effective support for upper-level service, and make the service products only give attention to the special service logics instead of to the general logics.

As an independent product, OP is developed independently. By absorbing, accumulating and implementing the precious experiences of other service products, OP has natural advantages of stability and high efficiency. It is a highly qualified platform for the service products.

Under the background that the company has had many production lines and has developed various products, the importance of OP platform has become much more notable. From the perspective of total investment and production, it can greatly reduce the development resource, improve the development efficiency and protect the product stability.

OP platform uses the JavaEE technology which is very mature and widely used in the industry, and it obeys the TMN and 3GPP regulations. It can be divided into 4 subsystems based on function coverage.

(1) UFP (Universal Foundation Platform) subsystem (basic service, scalable protection, specification protection, stability, HA protection, etc.);

(2) B-UFP (Business Universal Foundation Platform) subsystem (business modeling, basic business process, business service, etc.);

(3) UIP (User Interface Platform) subsystem (UI development structure, various group package, etc.);

(4) GBM (General Business Module) subsystem (general products for the end user, such as system supervision, safety management, and log system).

2) OP-BA

OP-BA provides the following functions:

(1) supervision among systems;

(2) supervision in the system;

(3) router management;

(4) work/service flow management;

(5) task management;

(6) patch management.

3) User end

The user end is the main part for human-machine interaction of OMC. The user can conveniently operate the OMC system from the user end, which can satisfy the equipment management needs. The user end relies on separation design:

(1) the separation of OMC system login from service system login;

(2) system login separation for various service.

The user has to log onto different service systems after he has logged onto the OMC system to meet management requirements of different services.

4) Server

The server is the core of the OMC system management. Its functions include the following parts.

(1) user access management: control the number of user access;

(2) user service management: provide service to user, such as, warning and reported table;

(3) equipment access management: control the equipment access;

(4) equipment service management: complete the equipment data report or setting, etc.

5) Interface with the outside

The design of the interface with the outside also obeys the designing principle of the server. Based on the workflow of platform, it can segment the service and customize the related process. Therefore, it can satisfy the frequent change of service.

4. Technical specification

1) Technical specification of the system

According to the network size, OMC can provide various kinds of setting plans. Different setting plan has different technical specifications and requirements for the hardware type. Table 2-3-1 shows the technical specifications of OMC.

Table 2-3-1　Technical specifications of the LTE-OMC system

Capacity specification	Specification value					
Setting	Super capacity		Large capacity		Normal	
Disk array capacity	3.6 T		3.6 T		3.6 T/1.7 T	
Server model	SUN NetraT5440	SUN FireV890	SUN NetraT5440	SUN FireV890	SUN NetraT5440	SUN FireV890
RAM	64 G	32 G	32 G	32 G	32 G	16 G
Possible managed area	30 000		18 000		9000	
Possible managed base station	3000		2000		1000	

2) Performance specification

The processing capacity of OMC is mainly reflected on processing a huge amount of warnings, events and specifications. The frequency and size of the warning, event and specification reported by each network element are as follows.

(1) LTE access network:

① the warning frequency reported by eNode B: <100/day;

② the event frequency reported by eNode B: <30/day;

③ the eNode B specification data: each area reports the specification data every 15 minutes: 2000 kb;

④ Each eNode B contains 3 areas.

(2) EPC network element:

① the warning frequency reported by each network element of EPC: <100/day;

② the event frequency reported by each network element of EPC: <30/day;

③ the specification data of each network element of EPC: report the specification data every 15 minutes: 2000 kb.

(3) IMS network element:

① the warning frequency reported by each network element of IMS: <100/day;

② the event frequency reported by each network element of IMS: <30/day;

③ the specification data of each network element of IMS: report the specification data every 15 minutes: 2000 kb.

3) Reliability

(1) In order to successively provide the operation and maintenance service for a long time, OMC system supports the followings:

① mean time before failure (MTBF): >200 000 hours;

② mean time to failure (MTTF): <30 minutes.

(2) If the user has high requirements for usage, OMC can choose the following methods to satisfy the user requirements.

① universal server uses 1+1 load sharing.

② business server uses n+1 deployment.

③ network element adapter uses n+1 deployment.

(3) If any of the business server or the network element adapter fails to provide service, the process cannot be recovered, it can quickly switch to the backup equipment to continue the work.

The related designing function is:

① management domain setting;

② system process supervision.

4) Other specifications

(1) OMC system can store the specification data and the warning data for 3 months, and the log of user operation for 12 months at least.

(2) The data synchronization time of a single network element is <=3 min.

(3) The processing time of the specification data file at a time is <=5 min.

(4) The displaying time delay of the warning reported to the terminal is less than 5 seconds.

(5) Tasks of the signal order tracking for each network element is no more than 16.

(6) The maximum creation number of the threshold for each network element specification is 32.

(7) The maximum number for user terminals is 128.

Training Module 3: Base Station Starting and Testing (20 Class Hours)

【Basic description】

For base station starting and testing, Datang Mobile Communication Equipment Limited Company has developed a software local operation and maintenance terminal (LMT). Its main functions include configuration management of the network element of the base station, fault management, performance measurement, safety management, etc. The construction and starting of a base station can be divided into base station starting preparation, software and hardware updating, board planning, local area establishment and area checking. The project is designed as the typical work task for the workers who enters the field of base station starting for the first time in the mobile communication system, and the training purpose is to familiarize the starting process of a base station. The training includes software and hardware preparation, LMT installation, environment construction, equipment updating, board planning, local area planning, area parameter setting and common failure resolution, etc. Through the training of the project, the learners can meet the basic requirements for starting and testing the mobile network of a base station. For a newly established station, the starting and testing process is shown in Figure 3-1-1.

【Training Purposes】

1. Knowledge to be known

(1) Master the starting process for a mobile communication base station.

(2) Master LMT software installation and setting.

(3) Master the common failure solutions for base station starting.

2. Ability to be learned

(1) Be familiar with the use of LMT software.

(2) Master the starting and testing methods.

(3) Solve the common starting failures.

【Training Requirements】

1. Preparation for tools, instruments and equipment

2. Knowledge evaluation points

(1) Board planning.

(2) Local area planning.

(3) Area parameters setting.

3. Skill evaluation points

(1) Complete the environment construction.

(2) Master the software and hardware updating for RRU.

(3) Solve the common failures.

```
                                    ┌─────────────┐
                                    │  starting   │
                                    └──────┬──────┘
┌────────────────────────┐                ▼
│  Hardware preparation  │────┐    ┌──────────────┐
├────────────────────────┤    │    │Preparation before│
│  Software preparation  │────┤    │staring       │
├────────────────────────┤    ├───▶└──────┬───────┘
│Installation,operation and│  │           ▼
│maintenances tools      │────┤    ┌──────────────┐
├────────────────────────┤    │    │  Equipment   │
│    Normal testing      │────┘    │  preparation │
└────────────────────────┘         └──────┬───────┘
                                          ▼
┌────────────────────────┐         ┌──────────────┐
│ PC sets IP router setting│───┐    │LMT installment│
└────────────────────────┘   │    └──────┬───────┘
                             ├──────────▼
┌────────────────────────┐   │    ┌──────────────┐
│LMT installment to connect│──┘    │Network planning│
│the base station        │         └──────┬───────┘
└────────────────────────┘                ▼
```

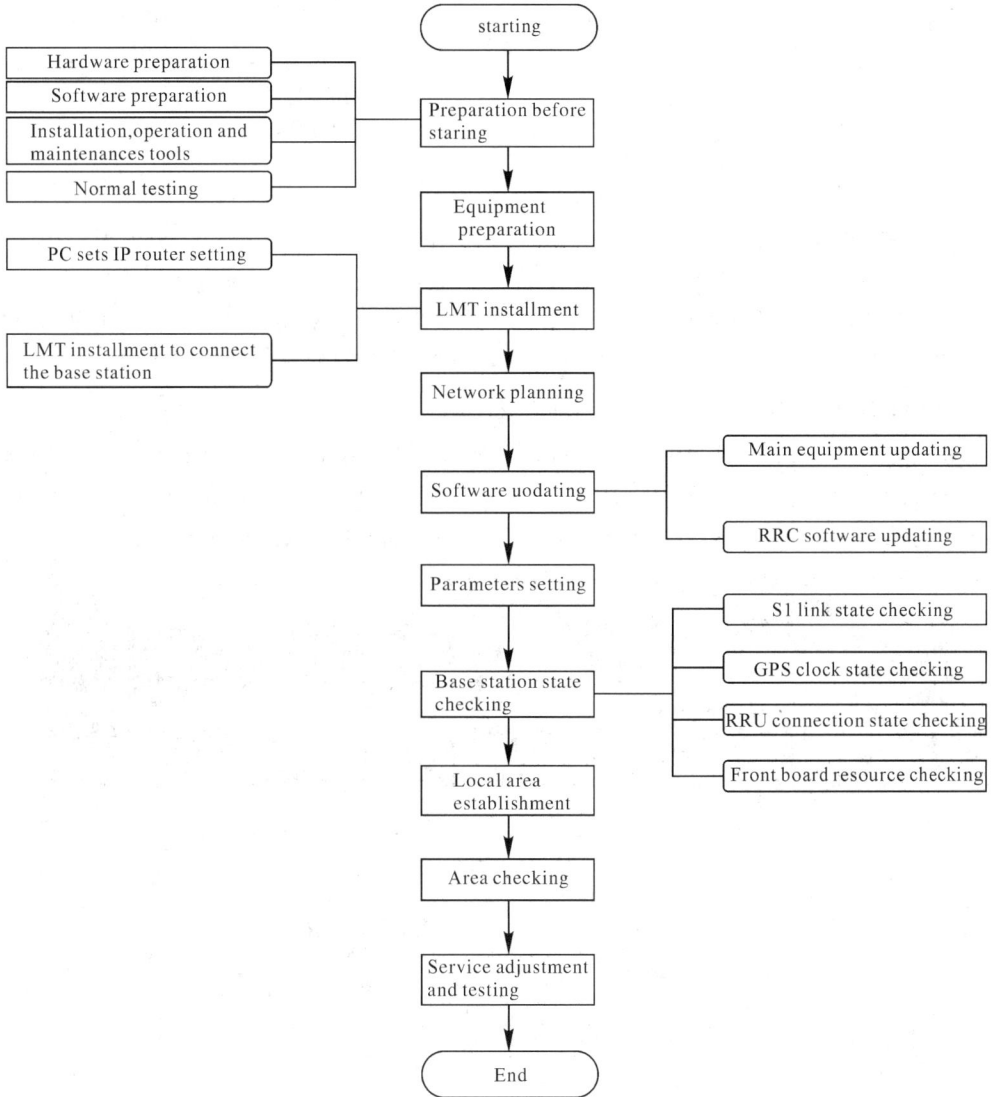

Figure 3-1-1   Base station starting & testing

Work Task 1: Preparation before Base Station Starting

1. Hardware preparation

Install the required equipment, as shown in Table 3-1-1.

Table 3-1-1   Installation equipment list

Hardware name	Number	Hardware name	Number
PC	1	EMA	1
EMB5116 case	1	HUB	1
SCTE single board	1	331FAERRU	1
BPOH single board	1	Net cable	Several
PSU single board	1	6G optical module	Several
FCU single board	1	Attenuator, bulkhead	Several

2. Software preparation

It is necessary to prepare the main station software package 5116TDL_V3.20.00.45.21.DTZ, RRU software package DTRRU_V3.20.00.45.21.DTZ, tool software LMT_LTE.rar and ATP_LTE.rar. General speaking, the authorized software package includes all the above softwares. The user can directly use the related softwares from the software package.

3. Environment construction

The single board can be plugged into the front panel, just as shown in Figure 3-1-2, and all the connection lines can be connected according to the setting environment shown in Figure 3-1-3. Turn on the EMB5116 base station and RRU when all the tasks are finished.

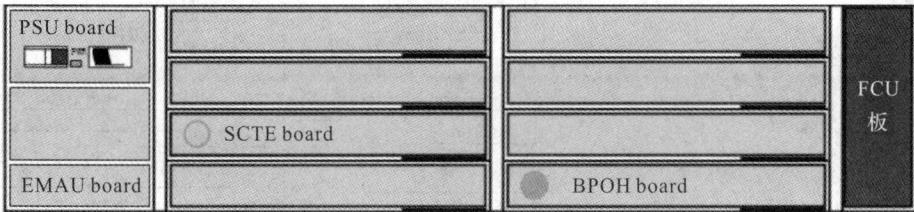

Figure 3-1-2   Front panel of EMB5116 base station

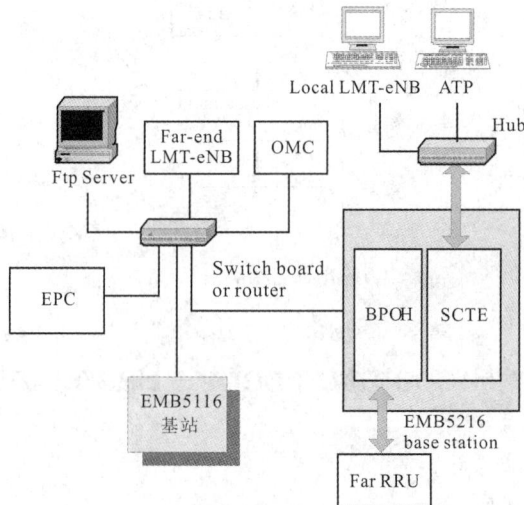

Figure 3-1-3   Setting environment

4. LMT installation

Find the LMT_LTE.rar in the authorized software package, then decompress it, open the LTE-LMT file, choose setup.exe and click to install LMT, choose the next button until the installation is finished. You can find the LMT icon on the desk when the installation is done. The details are as follows.

(1) Click LMT installation package setup.exe, as shown in Figure 3-1-4.

Figure 3-1-4　installation software

(2) Enter the installation interface, as shown in Figure 3-1-5.

Figure 3-1-5　Installation interface1

(3) Click the Next(N) button, as shown in Figure 3-1-6.

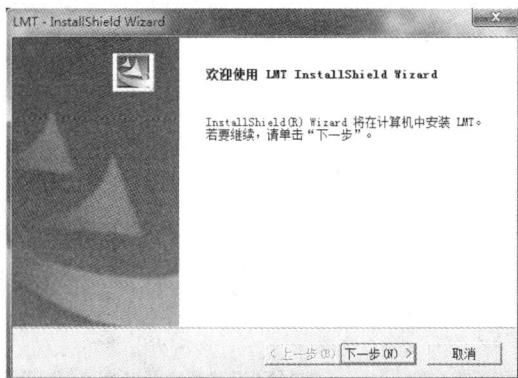

Figure 3-1-6　Installation interface2

(4) Fill in the user name and company name (you can choose what you like), and click the Next(N) button, as shown in Figure 3-1-7.

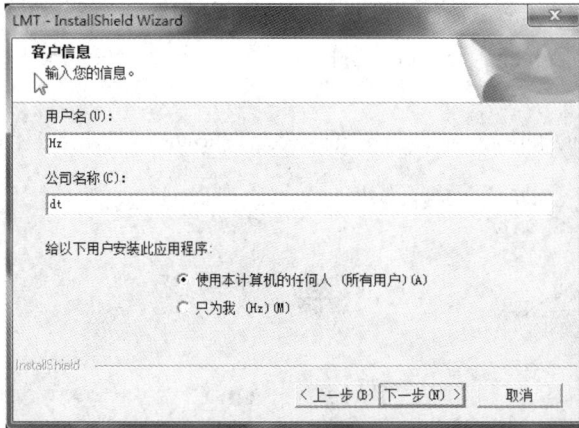

Figure 3-1-7　User information input interface

(5) Choose the 'typical (T) ' and click the 'Next(N)' button, as shown in Figure 3-1-8.

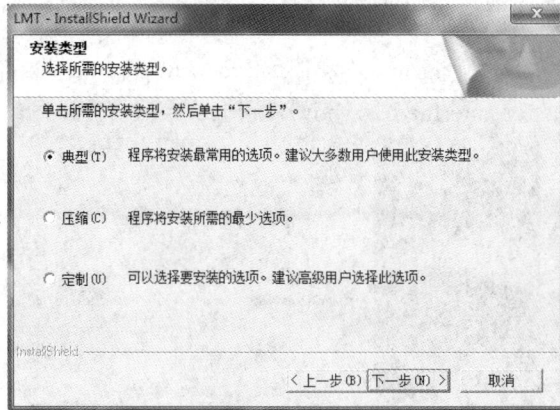

Figure 3-1-8　Installation type interface

(6) Enter the installing state, as shown in Figure 3-1-9.

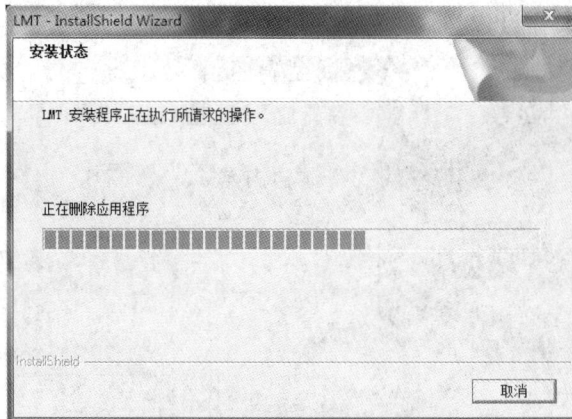

Figure 3-1-9　Installation process interface

(7) The installation indication for the public network access to the system software will appear after the installation completion in several seconds, and just click the 'Confirm'(确定) button, as shown in Figure 3-1-10.

(8) Click the 'Next(N)' button to enter the next step, as shown in Figure 3-1-11.

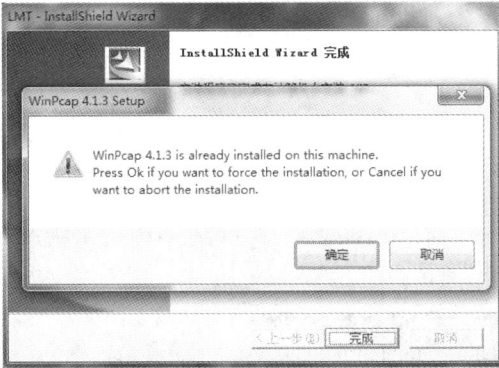

Figure 3-1-10　Installation indication interface

Figure 3-1-11　Installation interface of WinPcap

(9) Choose the 'I Agree' button to agree the license, as shown in Figure 3-1-12.

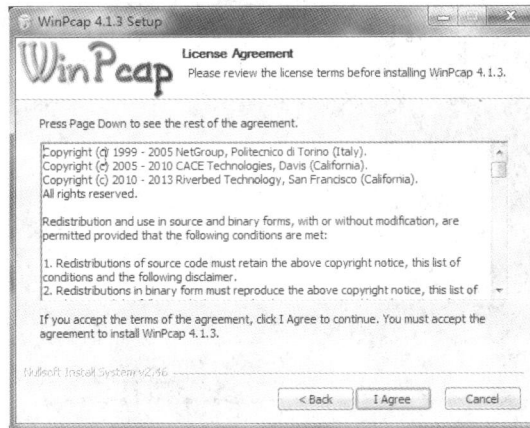

Figure 3-1-12　Installation agreement interface

(10) Make the default tick and install, as shown in Figure 3-1-13.

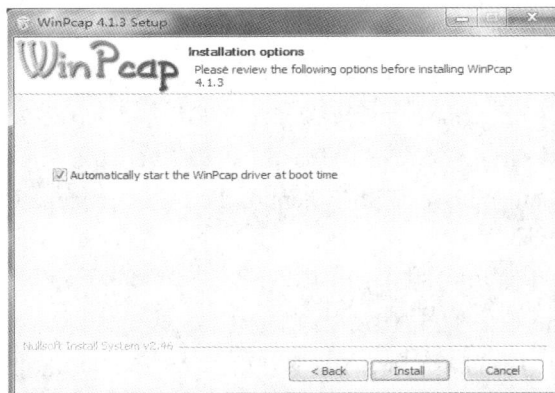

Figure 3-1-13　Default installation interface

(11) Click the 'Finish' button to end the installation, as shown in Figure 3-1-14.

Figure 3-1-14   WinPcap installation completion interface

(12) LMT installation has been finished, as shown in Figure 3-1-15. (If your system is not XP, you need to right-click the desk icon to use the administrator ID to login).

Figure 3-1-15   LMT installation completion interface

5. IP address setting

It is necessary to set IP address to the PC testing network card. Because the local testing IP address for SCT (Switch Control Transmission) is 172.27.245.92 and 10.10.1.192, SCT service interface IP address is 10.0.1.192, so the PC and SCT should be set to the same network segment. PC sets 172.27.245.233 with a mask code of 255.255.255.0, and 10.10.1.100 with a mask code of 255.255.0.0. It is necessary to add the 10.0 and the 10.1 network segment routers to the DOS system of the PC. Input "cmd" in the DOS system of the PC Start Menu Running, and the setting should be as follows:

route add 10.0.0.0 mask 255.255.0.0 10.10.1.192 -p

route add 10.1.0.0 mask 255.255.0.0 10.10.1.192 -p

The normal default IP address for RRU is 172.27.45.250, so the PC network card needs another IP address 172.27.45.100, and the mask code is 255.255.255.0. This IP address will be needed for later RRU update when the local maintenance tool logins into RRU.

(1) Click the network connection icon at the lower right corner to open the network and the sharing center, and right-click the local network connection to choose 'property(R)', as shown in Figure 3-1-16.

(2) Choose the following IP address and input the same network segment IP address 172.27.245.233as that of LMT maintenance interface, the subnet mask code is 255.255.255.0, and DNS (Domain Name System) can be empty. Click the 'Advanced(V)' button to input several IP addresses and enter the next adding process, as shown in Figure 3-1-17.

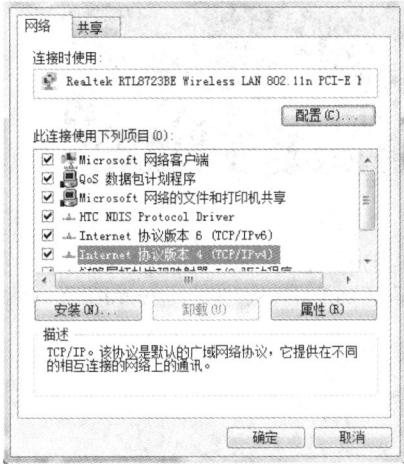

Figure 3-1-16   Network property interface

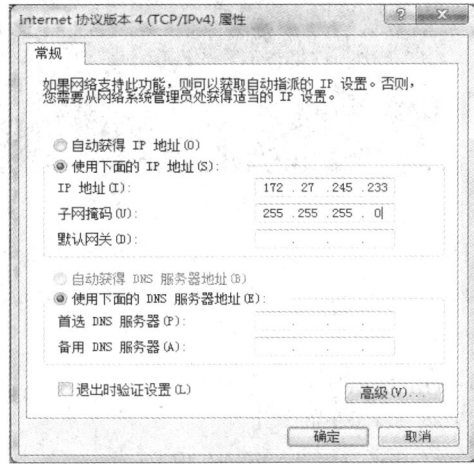

Figure 3-1-17   IP address setting interface

(3) You can add several IP addresses here, click the IP 'Address(A)' adding and enter the next step, as shown in Figure 3-1-18.

(4) Set the IP address for the equipment you want to add. For example, if it is desired to login into the RRU IP address 172.27.45.250, we need to add the same network segment IP address(I), as shown in Figure 3-1-19.

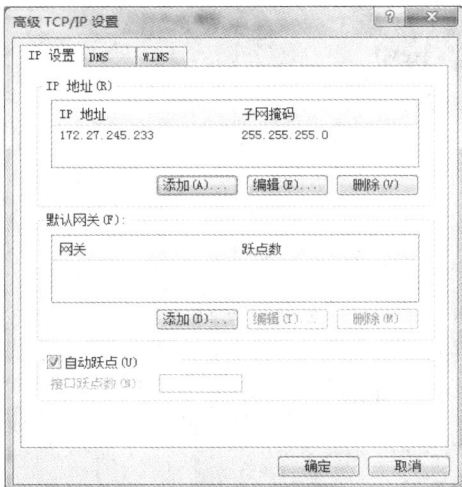

Figure 3-1-18   IP address advanced interface

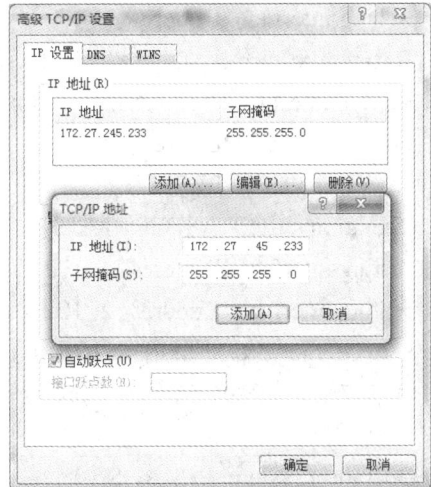

Figure 3-1-19   IP address addition interface

The adding of the testing interface IP address 2 and the service interface IP address is as follows.

Open the PC Start Menu, find the command prompt in the all-program-appendix(it is required to input the administrator IP to open it in Win7 system, and add what is shown in Figure 3-1-20.

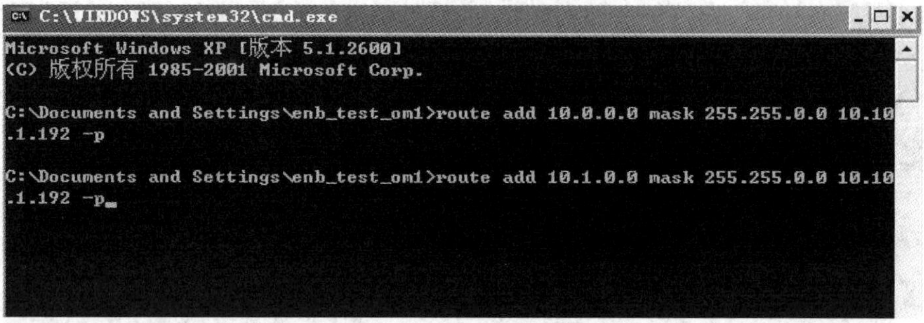

Figure 3-1-20   Router adding interface.

The IP address settings for various boards are as follows:

The IP address planning for EMB5116 hardware platform and software operation system is as follows:

(1) 0/1 slot SCT:

Processer: 8548 SCP

Operation system: VXWORKS

Testing interface IP address 1: 172.27.245. (91+slot number)

Testing interface IP address 2: 10.10.slot number.192

Service interface IP address: 10.0.slot number.192

(2) 2–7 slots BPOE core 0:

Processer: P2020 BCPE0

Operation system: LINUX

Testing interface IP address 1: 172.27.246.(1+slot number)

Testing interface IP address 2: 10.10.slot number.192

Service interface IP address: 10.0.slot number.192

(3) 2–7 slots BPOE core 1:

Processer: P2020 BCPE1

Operation system: VXWORKS

Testing interface IP address 1: 172.27.246.(21+slot number)

Testing interface IP address 2: 10.10.slot number.193

Service interface IP address: 10.0.slot number.193

(4) 2–7 slots BPOE DSP (Digital Signal Processing) core:

Processer: 8156 DSP

Operation system: PHAROS

Testing interface IP address: 10.10.slot number. (192+ processer PROCID)

Testing interface IP address: 10.0.slot number. (192+ processer PROCID)

Processer PROCID range: 3–20

The ID of the DSP processer starts from 3 and ends with 20 (there are 18 cores in total).

The PROCID numbering rule of DSP processer is shown in Table 3-1-2.

Table 3-1-2　numbering rule

	Core 0	Core 1	Core 2	Core 3	Core 4	Core 5
DSP0	3	4	5	6	7	8
DSP1	9	10	11	12	13	14
DSP2	15	16	17	18	19	20

(5) RRU:

Processor: 8313

Operation system: LINUX

Testing interface IP address 1: 172.27.45.250

Service interface IP address: 10.1.32+RRU number.192

Take the 4-slot BPOE as an example for the DSP address, which is shown in Tables 3-1-3, 3-1-4 and 3-1-5.

Table 3-1-3　Core0 local testing IP address

Slot number	Board name	Core0 local testing IP address	Core0 service IP address
0	SCT	172.27.245.91	10.0.0.192
1	SCT	172.27.245.92	10.0.1.192
2	BPOE	172.27.246.3	10.0.2.192
3	BPOE	172.27.246.4	10.0.3.192
4	BPOE	172.27.246.5	10.0.4.192
5	BPOE	172.27.246.6	10.0.5.192
6	BPOE	172.27.246.7	10.0.6.192
7	BPOE	172.27.246.8	10.0.7.192

Table 3-1-4　Core1 local testing IP address

Slot number	Board name	Core1 local testing IP address	Core1 service IP address
0	SCT	—	—
1	SCT	—	—
2	BPOE	172.27.246.23	10.0.2.193
3	BPOE	172.27.246.24	10.0.3.193
4	BPOE	172.27.246.25	10.0.4.193
5	BPOE	172.27.246.26	10.0.5.193
6	BPOE	172.27.246.27	10.0.6.193
7	BPOE	172.27.246.28	10.0.7.193

Table 3-1-5   PROCID local testing IP address

DSP processor PROCID	Local testing IP address	Service IP address
dsp-core0	10.10.4.195	10.0.4.195
dsp-core1	10.10.4.196	10.0.4.196
dsp-core2	10.10.4.197	10.0.4.197
dsp-core3	10.10.4.198	10.0.4.198
dsp-core4	10.10.4.199	10.0.4.199
dsp-core5	10.10.4.200	10.0.4.200
dsp-core6	10.10.4.201	10.0.4.201
dsp-core7	10.10.4.202	10.0.4.202
dsp-core8	10.10.4.203	10.0.4.203
dsp-core9	10.10.4.204	10.0.4.204
dsp-core10	10.10.4.205	10.0.4.205
dsp-core11	10.10.4.206	10.0.4.206
dsp-core12	10.10.4.207	10.0.4.207
dsp-core13	10.10.4.208	10.0.4.208
dsp-core14	10.10.4.209	10.0.4.209
dsp-core15	10.10.4.210	10.0.4.210
dsp-core16	10.10.4.211	10.0.4.211
dsp-core17	10.10.4.212	10.0.4.212

Work Task 2: Normal Update for Software and Hardware of Base Station

1. The steps for LMT to login onto a base station

(1) Click the LMT icon on the desk (Win7 system needs the administrator authentication) after its installation and enter the LMT operation interface, as shown in Figure 3-2-1. The default user name is 'administrator', the password is '111111', and choice style is 'LMT'. After entering the LMT operation interface, a LMTAgent will pop-up at the same time, then you need to choose the monitor network board in the network board list, and click the confirm button, as shown in Figure 3-2-2.

(2) FTPServer software will pop-up after login (do not shut down the software, which is essential for the data interaction between the PC used and the base station EMB5116, and also it influences the base station login), the server log interface is shown as in Figure 3-2-3.

Figure 3-2-1   LMT login interface

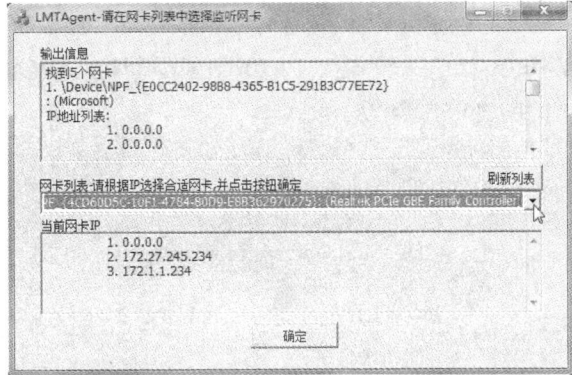

Figure 3-2-2   Choosing network board interface

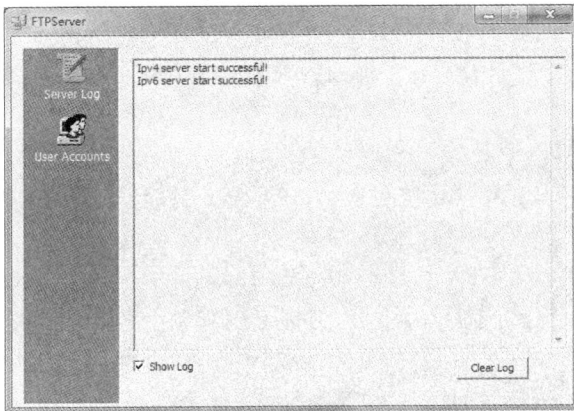

Figure 3-2-3   FTPServer log interface

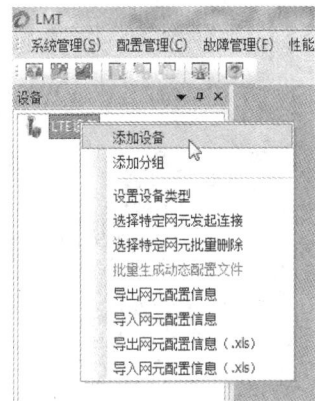

Figure 3-2-4   Network element
adding interface

(3) Add new network element, as shown in 3-2-4.

(4) Choose the ENODEB as the network element type, give the network element friendly name according to the real situation, and then confirm. There will be a base station icon, which was just set under the LTE equipment, as shown in Figure 3-2-5.

(5) Right-click the newly added network element base station, and choose setting, as shown in Figure 3-2-6.

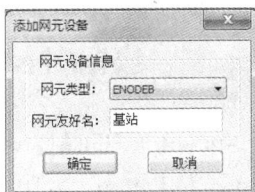

Figure 3-2-5   Network element choosing interface

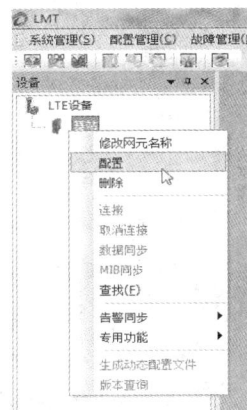

Figure 3-2-6   Network element operation option

(6) Choose the setting in the pop-up dialog to set the network element, and the IP is the SCT IP, 172.27.245.92 or 10.10.1.192. Others are default and then you click to confirm, as shown in Figure 3-2-7.

(7) Right-click the network element base station, the dialog box will pop-up and choose 'connect'. The LMT starts to connect, as shown in Figure 3-2-8.

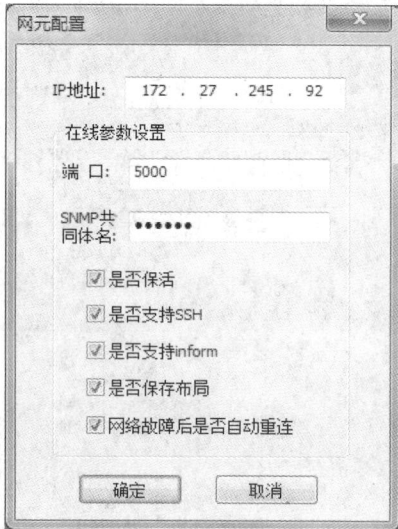

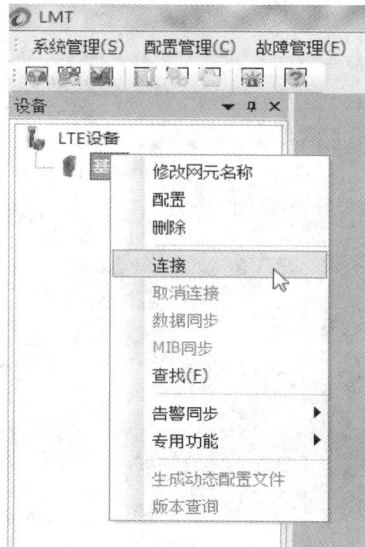

Figure 3-2-7　Network element setting interface　　Figure 3-2-8　Base station connection option

(8) Look at the pop-up message in the information-browsing page, the login will be successful when the received result event of the file upload/download and the data consistency processing are the same, as shown in Figure 3-2-9.

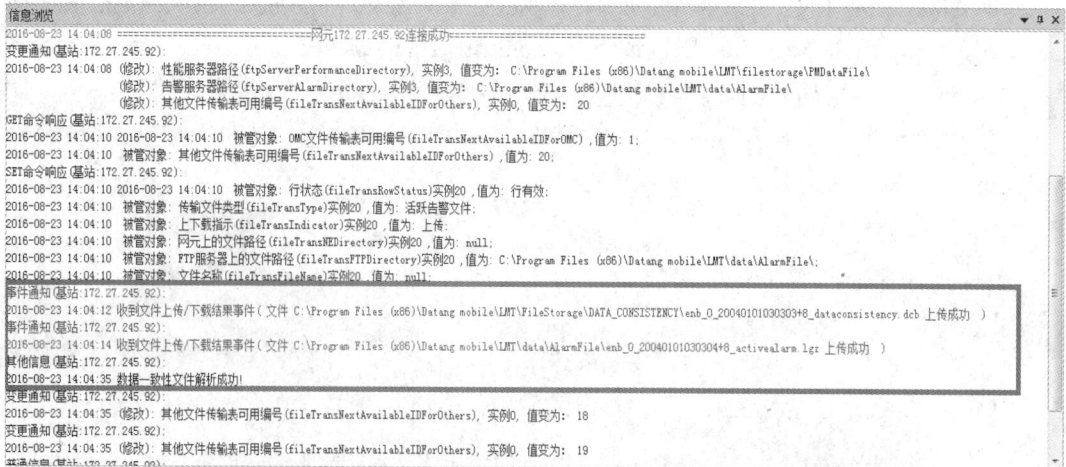

Figure 3-2-9　Information browsing page

2. Software and hardware update for the radio frequency unit RRU

(1) Click the file management in the LMT tool bar, as shown in Figure 3-2-10.

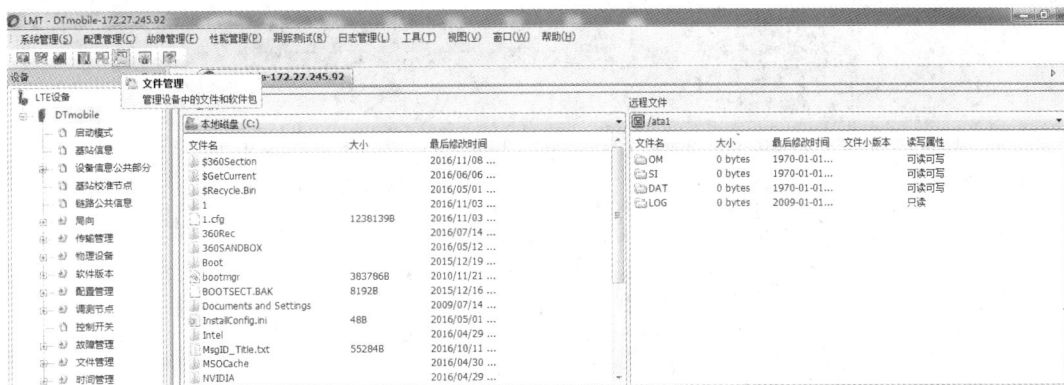

Figure 3-2-10   File management interface

(2) Choose the saving route of the RRU software package in the local file of LMT file management interface. 338FA and 348FA need to be configured. The extension point for 20M compression + 10M compression uses LTEV2RRU_kr.dtz, and other sites use LTEV2RRU_fkr.dtz, as shown in Figure 3-2-11.

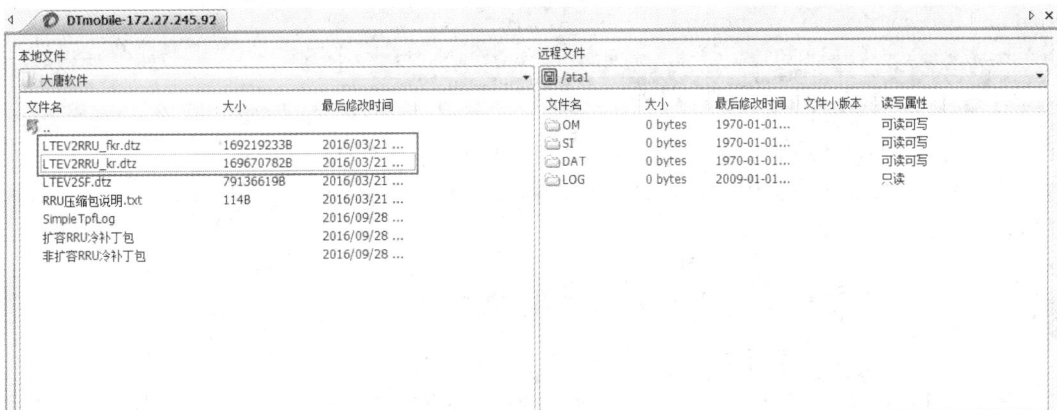

Figure 3-2-11   File saving route interface

(3) Directly drag the RRU software package into the base station on the right side. The indication will be given if the versions are the same, and there will be no indication if the versions are different. Then, we need to click the 'confirm' button, as shown in Figure 3-2-12.

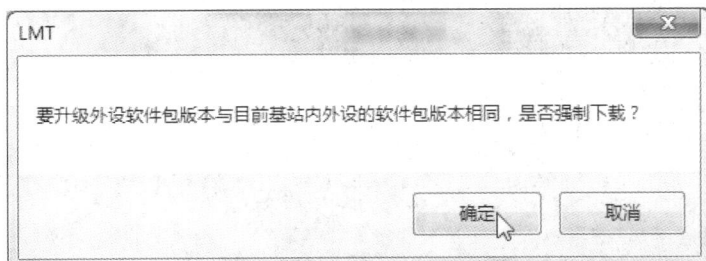

Figure 3-2-12   Compel download acknowledge

(4) The software package download and activation interface will pop-up, and click to confirm. The RRU software will be downloaded and activated, as shown in Figure 3-2-13.

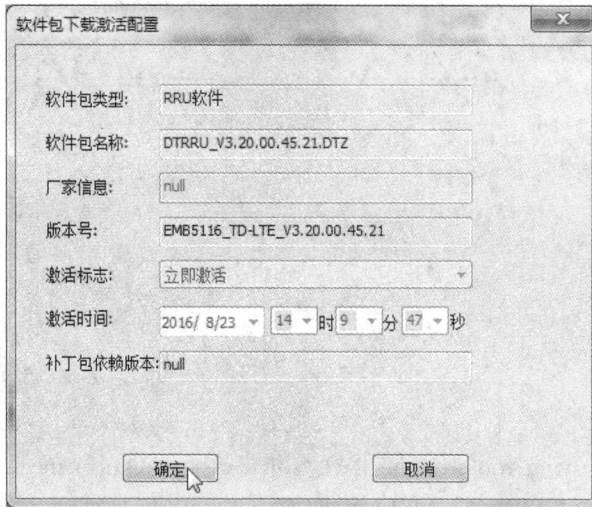

Figure 3-2-13　Download and activation interface

(5) Give attention to the file download progress, and it can be divided into two parts: down load and decompression, as shown in Figure 3-2-14.

Figure 3-2-14　Download progress interface

(6) After the software is downloaded, do nothing except decompress it, as shown in Figure 3-2-15.

Figure 3-2-15　Decompression process interface

(7) You need to give attention to another point that RRU has two cold patch packages, and you can choose to download it or not, as shown in Figure 3-2-16.

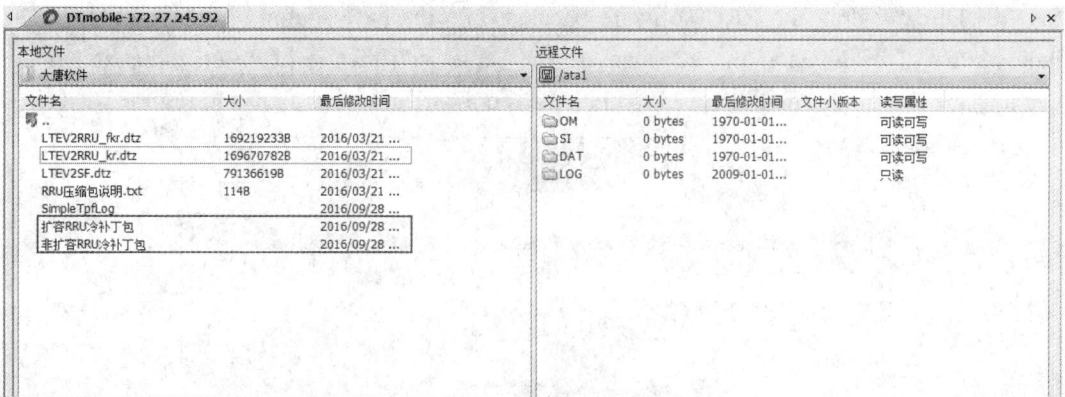

Figure 3-2-16　Cold patch package interface

Important notice: It is necessary to reset the base station manually in making the software effective if only the RRU would need to be updated. Otherwise, the RRU software package will not be complete and the version fallback mistake may happen (it is not necessary to reset manually if the main equipment software starts to update after the RRU has finished its updating, and the equipment can automatically reset itself after the main equipment software has finished updating).

The manual resetting steps go as follows.

(1) Figure 3-2-17 shows how to reset the equipment.

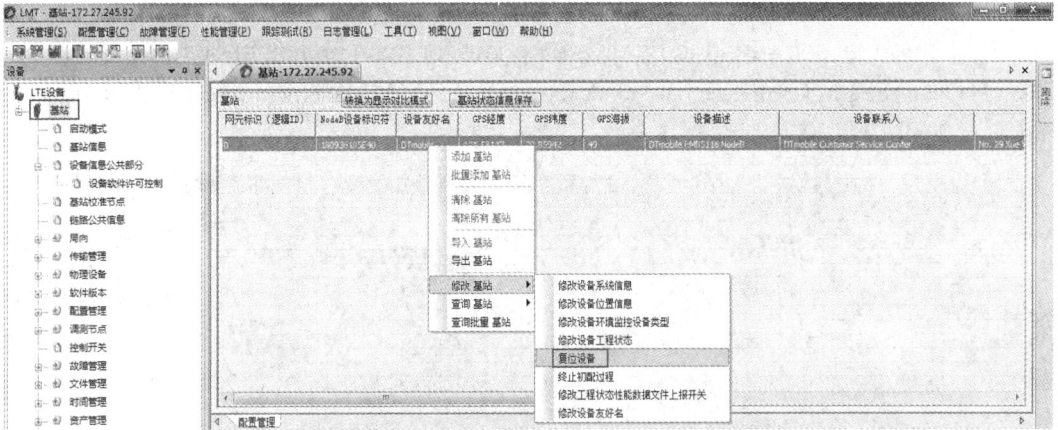

Figure 3-2-17   Reset equipment interface

(2) Decide whether it is necessary to produce the dynamic setting file, which is the resetting record. Choose to produce the dynamic setting file and resets it, as shown in Figure 3-2-18.

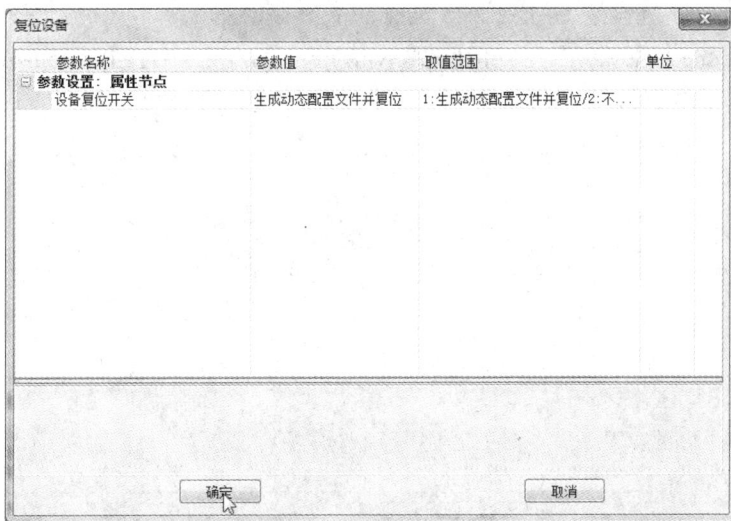

Figure 3-2-18   Reset choosing interface

(3) Wait a moment after confirmation, and then the base station will automatically disconnect. If the data compatibility processing file cannot be displayed, what needs to be done is to disconnect and reconnect the network element manually, or shut down and reopen the software (LMTAgent and FTPServer need to be shut down at the same time when the software is shut down), as shown in 3-2-19.

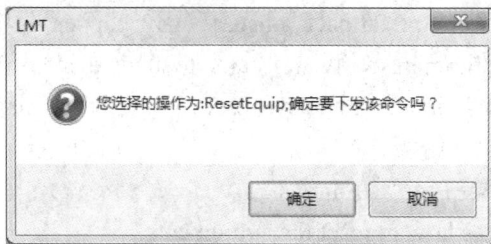

Figure 3-2-19　Order issuing interface

3. The normal software and hardware update for main equipment of a base station

(1) Open the LMT software, and click the icon of File Management in the LMT toolbar, as shown in Figure 3-2-20.

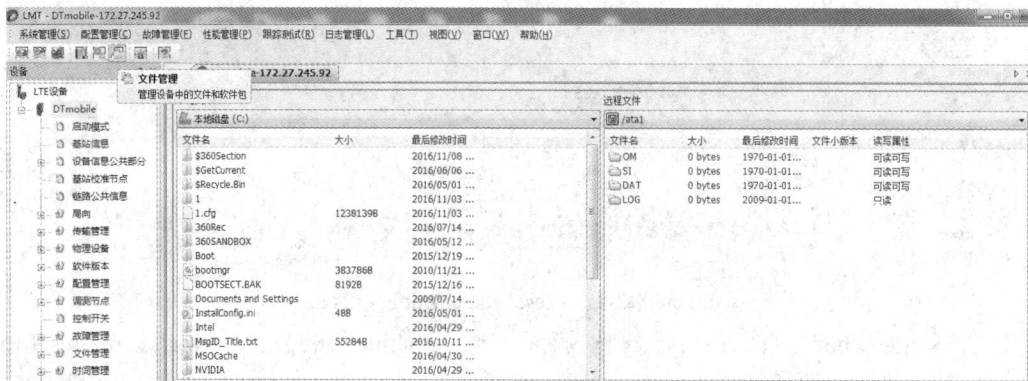

Figure 3-2-20　File management interface

(2) Choose the LTEV2SF.dtz software saving route in the LMT file management window, as shown in Figure 3-2-21.

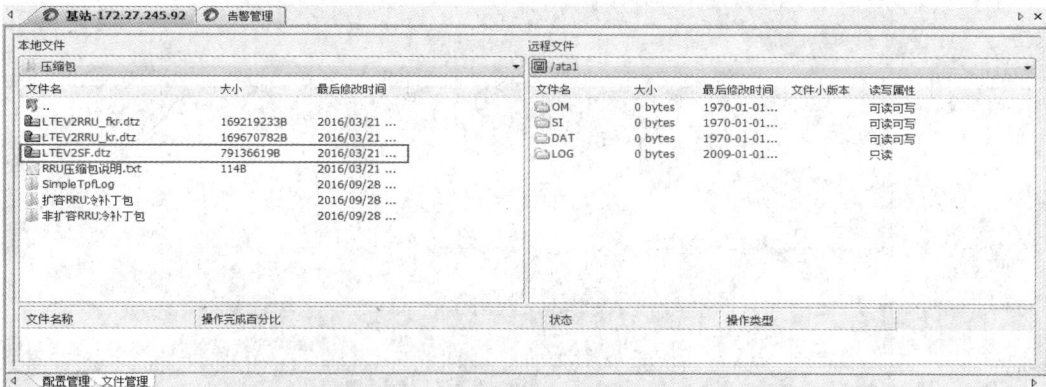

Figure 3-2-21　File choosing interface

(3) Directly drag the LTEV2SF.dtz software package into the base station on the right side. There will be an indication whether to download forcibly if the versions are the same, and there will be no indication if the versions are different. We can click to confirm and enter the download and activation setting interface of the software package. For the hardware activation indication, the 'activation mode' should be chosen, as shown in Figure 3-2-22.

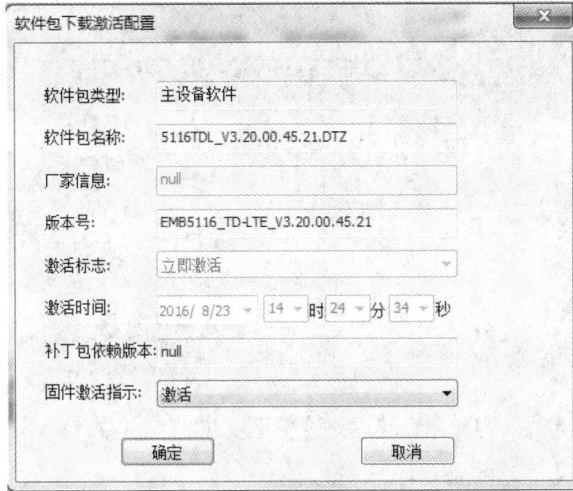

Figure 3-2-22   Download and activation interface

(4) We don't need to do any operation after confirmation so as to avoid the software package download break and download failure, etc. The base station can automatically disconnect and reconnect when the message dialog box pops up. If the data compatibility processing file cannot be displayed, what needs to be done is to disconnect and reconnect the network element manually, or shut down and reopen the software (it is necessary to shut down both LMTAgent and FTPServer at the same time when the software is shut down), as shown in 3-2-23.

Figure 3-2-23   Information-viewing page

4. Version check

The base station can automatically restart after successful software activation. By checking the file management, you can find out whether the present version is the version just updated after the base station restarting.

Use the mouse to right-click the remote file in the file management interface and choose the version checking in the pop-up window. You need to choose the base station software package version by the software package version or outer software package version for checking, as shown in Figure 3-2-24.

You can find out the present version information in the pop-up dialog box, as shown in Figure 3-2-25.

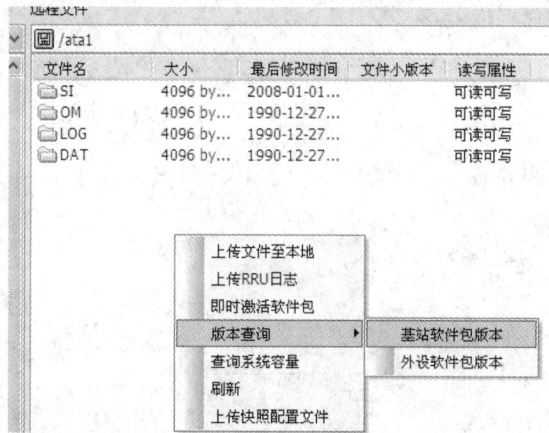

Figure 3-2-24　Version checking interface

Figure 3-2-25　Software version information

The RRU version update for EMB5116 and RRU can be carried out successfully according to the above instructions, which are the basic operation instructions for daily updates.

It is also possible to update the main station first and then you will find that the update of RRU cannot produce the dynamic setting file to reset the base station, and users usually prefer to use the gear initialization interface to reset base station. This resetting order of base station cannot produce the dynamic setting file to reset base station, and the saved version planning information for cur.cfg file is still the previous data. Therefore, it is often found that the RRU version fallback happens or the update is not successful after the base station restarting. In order to avoid this situation, it is needed to update RRU first and then the main station in the version update for the base station. Thus, the RRU version fallback will not happen again due to the resetting or power failure, etc.

Work Task 3: Board Planning

1. The graphic setting interface for new board planning

(1) Right-click the root directory of the network element equipment of the object tree, and choose the 'network planning' in the 'special function' and enter the board planning and setting interface, as shown in Figures 3-3-1 and 3-3-2(note: if the color for the network planning is gray and cannot be chosen, then go to Work Task 6 for the related reason and solution).

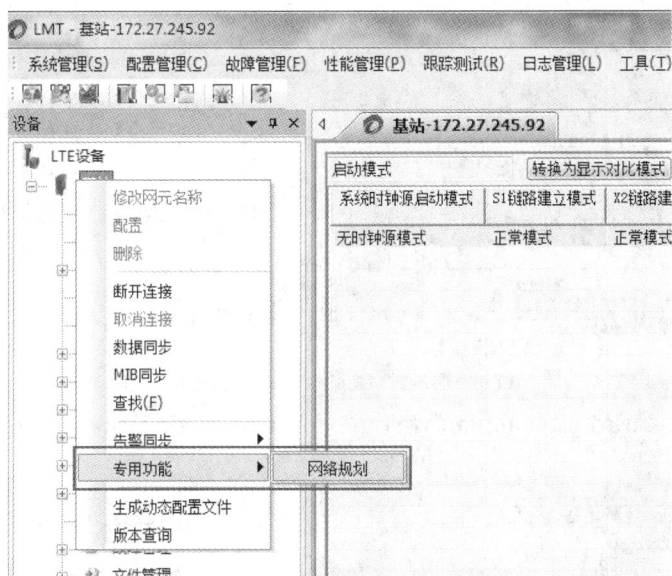

Figure 3-3-1　　Network planning interface

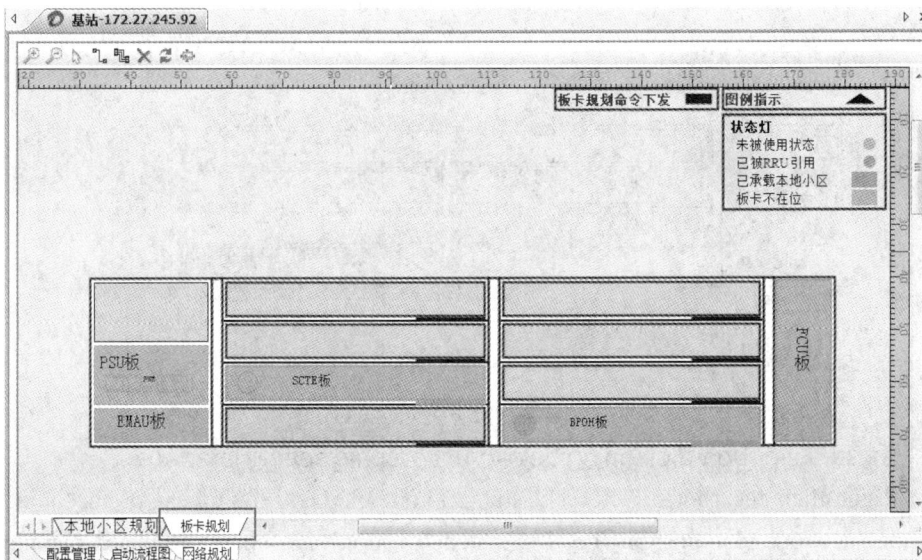

Figure 3-3-2　　Board planning interface

(2) Choose the related board which needs planning at the frame panel, double-click the empty place of the slot, then the board setting interface will pop up and choose the board type, and confirm it, as shown in Figure 3-3-3.

Figure 3-3-3　Board setting interface

(3) Click the 'board planning order' at the upper right corner to issue orders, as shown in Figure 3-3-4.

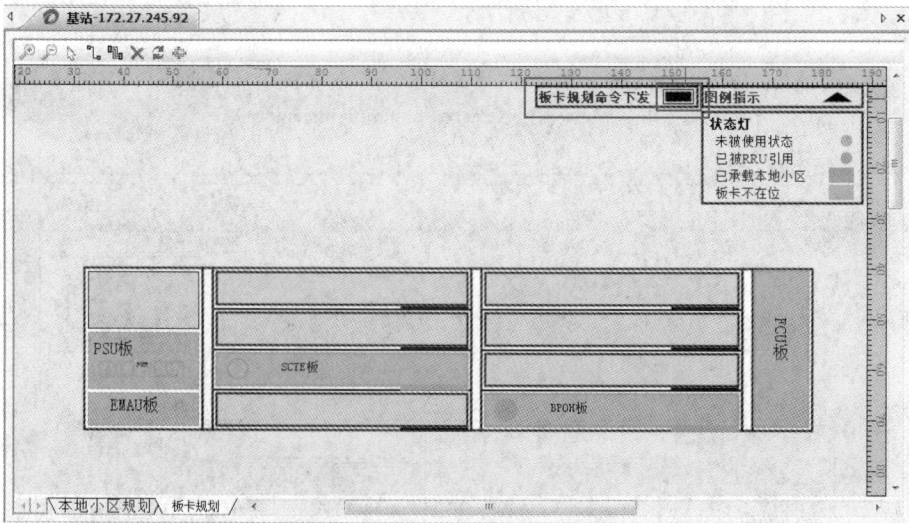

Figure 3-3-4　Oder issuing interface

2. The steps for board planning deletion of graphic setting interface

(1) Right-click the root directory of the network element equipment of the object tree, choose the 'network planning' in the 'Dedicated function' and enter the board planning and setting interface, as shown in Figures 3-3-5 and 3-3-6.

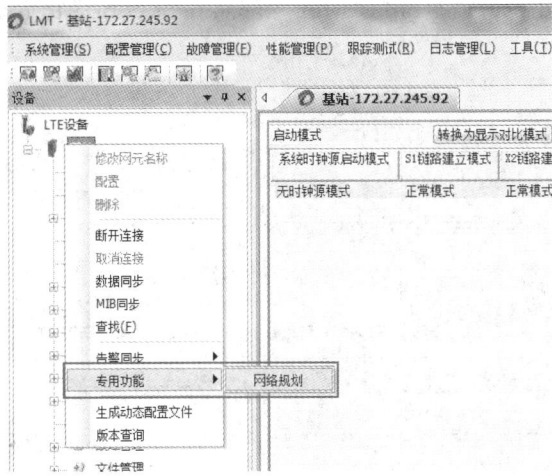

Figure 3-3-5　Network planning interface

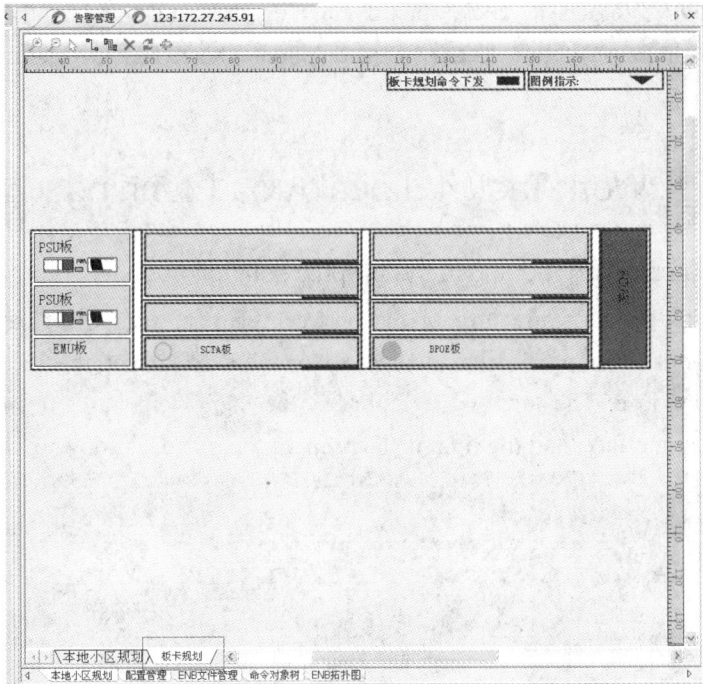

Figure 3-3-6　Board planning interface

(2) Choose the board whose planning needs to be deleted at the frame panel, and press the 'delete' key on your keyboard to delete, as shown in Figure 3-3-7.

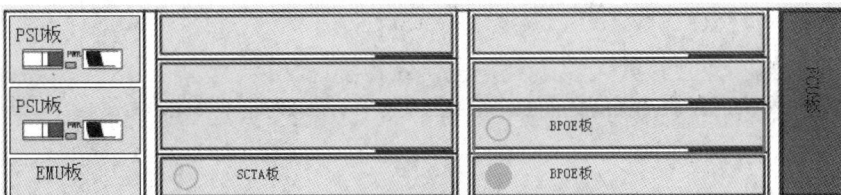

Figure 3-3-7　Board planning deletion interface

(3) Click the 'board planning order' at the upper right corner to issue orders, as shown in Figure 3-3-8.

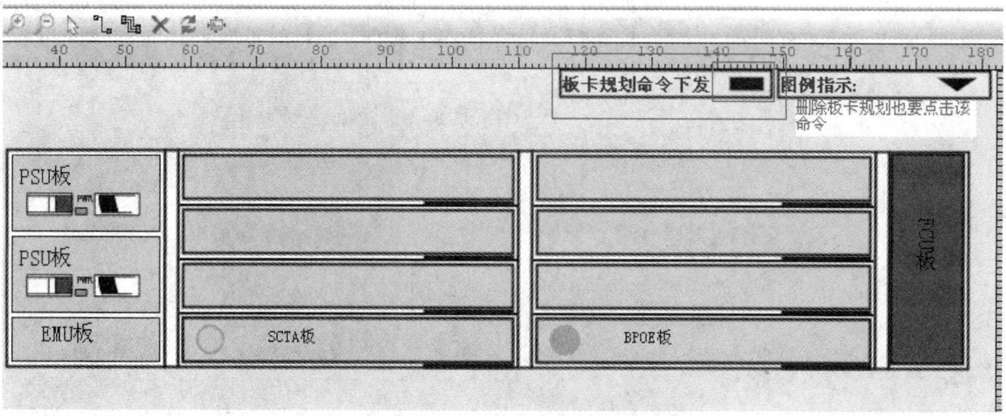

Figure 3-3-8　Order issuing interface

The user always forgets to click. 'board planning order to click' in the graphic setting interface, which can cause failure of the board deletion. Therefore, pay attention to this point during daily testing.

Work Task 4: Local Area Planning

1. Local area planning adding in the graphic setting interface

(1) Right-click the root directory of the network element equipment of the object tree, choose the 'network planning' in the 'dedicated function' and enter the local area planning graphic setting interface. The left of the Figure is the equipment base, in the middle is the planning operation interface, and the right is the property interface, as shown in Figure 3-4-1.

Figure 3-4-1　Local area planning introduction interface

(2) Right-click '0 Area' to start network planning in the pop-up window. You need to drag the elements you need to the middle operation interface, such as, the needed board, RRU, and antenna. BPOH should be put in the right slot number, You need to choose the right type of an antenna, which should be based on the actual requirements, for example, the working mode for RRU is the NORMAL mode as shown in Figures 3-4-2, 3-4-3, 3-4-4 and 3-4-5.

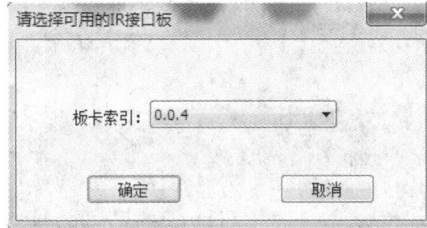

Figure 3-4-2   Interface board choosing interface

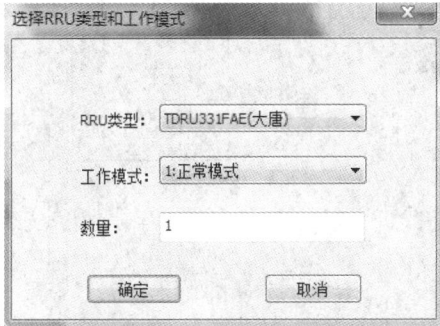

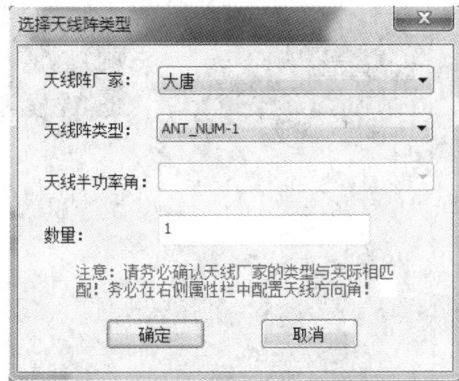

Figure 3-4-3   RRU work mode choosing interface   Figure 3-4-4   Antenna type choosing interface

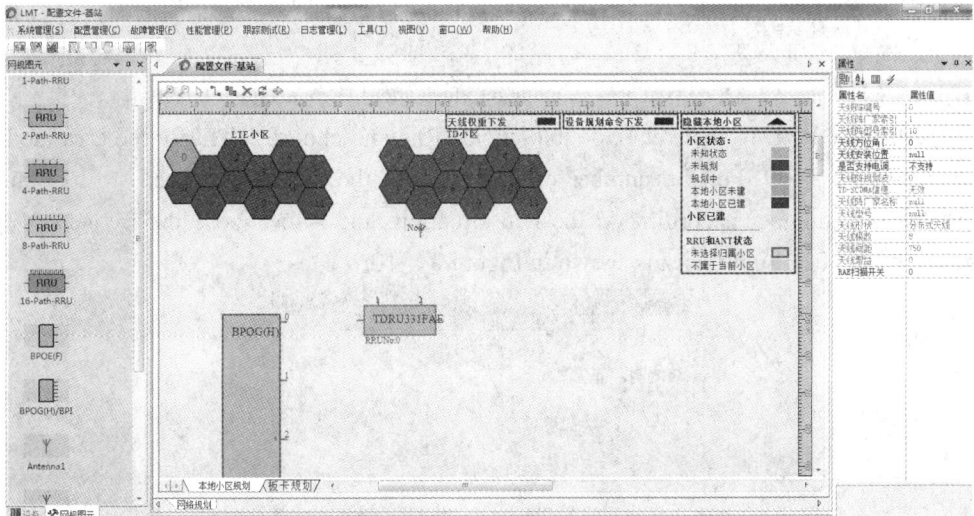

Figure 3-4-5   Local area planning interface after parameters change

(3) Start to connect the various elements after the preparation work is done. The connection mode interface is shown in Figure 3-4-6. You can click the 'single-connection line' or the

'multi-connection line' in the toolbar. The multi-connection can connect all the antenna interfaces with the RRU interface at the same time. For the rate of the optical module, choose '5G', as shown in Figure 3-4-7. Figure 3-4-8 shows the completion of the setting.

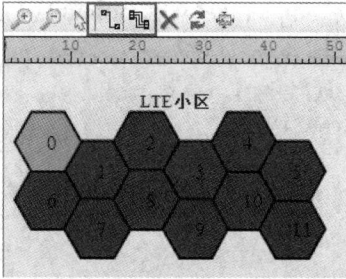

Figure 3-4-6   Connection mode interface

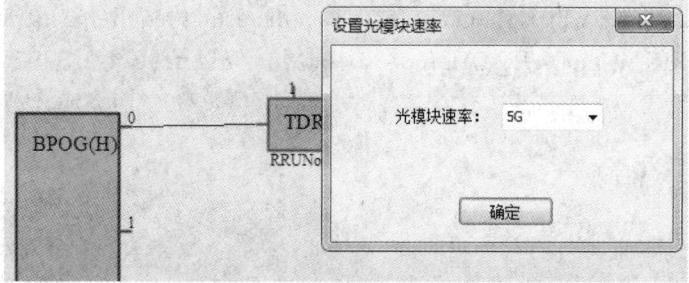

Figure 3-4-7   Optic module setting interface

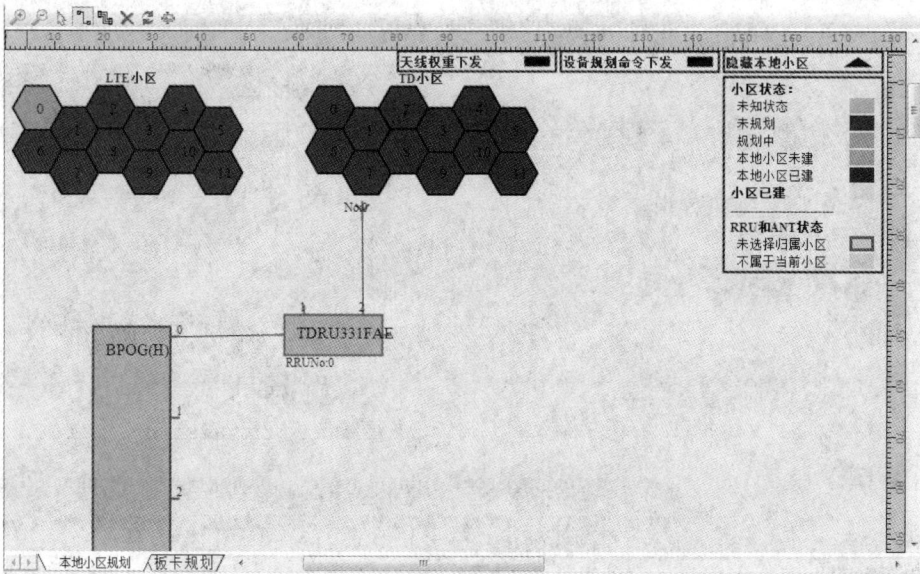

Figure 3-4-8   Local area planning interface after the connection of cables

(4) Attribute RRU to the area 0, double-click RRU, and choose 'LTE' for the area mode, as shown in Figure 3-4-9. For the number of LTE, choose the area which needs to be planned, choose the port number according to the real situation, and also choose the frequency range supported by the RRU interface, as shown in Figure 3-4-10.

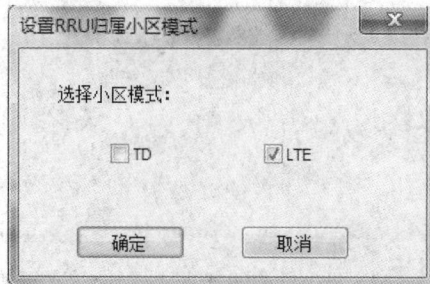

Figure 3-4-9   Area mode choosing interface

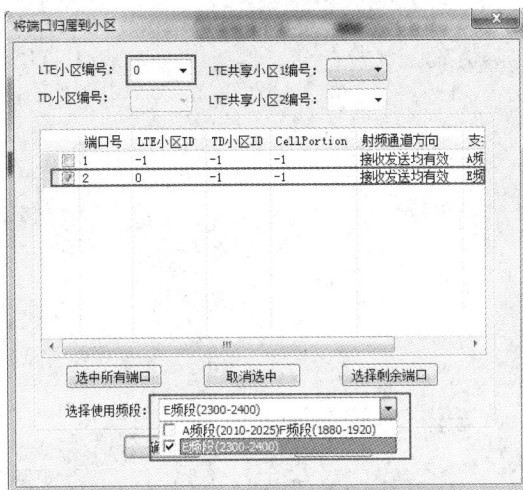

Figure 3-4-10   Area property choosing interface

(5) Modify the local area property value; click area 0, and modify the property value at the right of property parameter according to its actual value, as shown in Figure 3-4-11 and table 3-4-1.

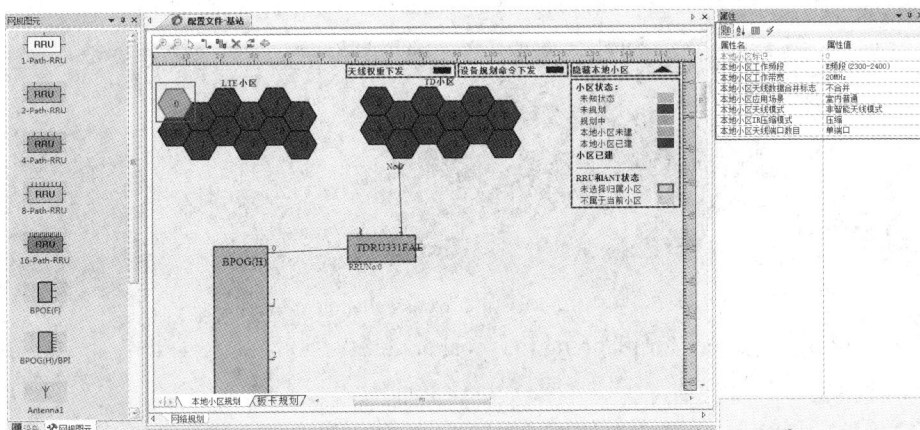

Figure 3-4-11   Local area planning property value modification interface

Table 3-4-1   Local area property parameter interface

Property name	Property value
Local area identification	0
Local area working frequency range	E frequency range (2300–2400)
Local area working bandwidth	20 MHz
Local area antenna data combination label	No combination
Local area application scene	Indoor normal
Local area antenna mode	Non-intelligent antenna mode
Local area IR compression mode	Compressed
Local area antenna port number	Single interface

(6) Click the 'board planning order', and right-click '0 area' to choose the 'network planning', as shown in Figures 3-4-12 and 3-4-13.

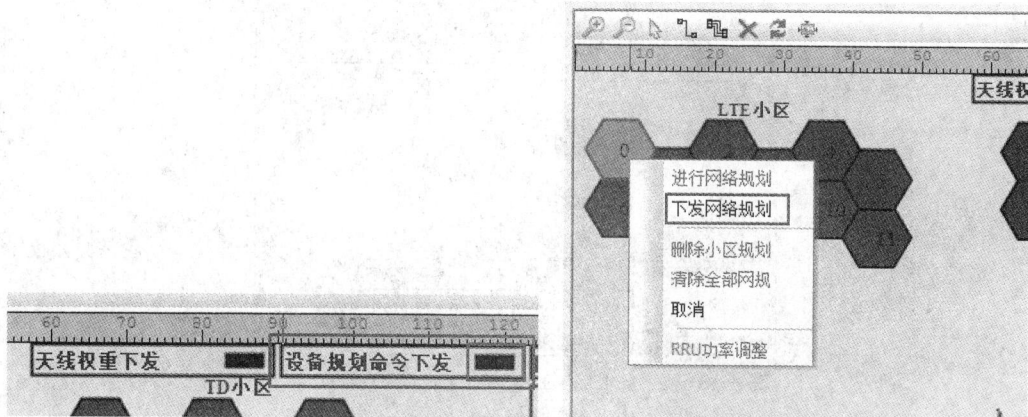

Figure 3-4-12   Board planning order issuing option   Figure 3-4-13   Network planning distribution option

The local area color will turn to ORANGE when the setting is finished. If RRU is connected and the local area is successfully established, then its color will turn to BLUE, as shown in Figure 3-4-14.

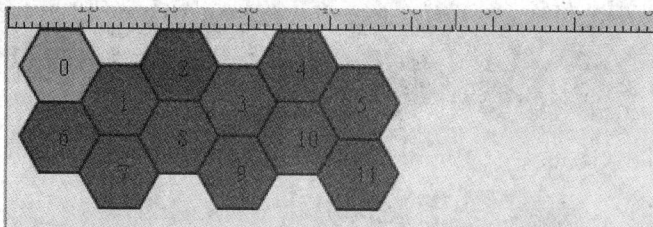

Figure 3-4-14   Successful area establishment interface

There will be no failure for planning the local area if the above instructions about local area planning are strictly followed.

2. Graphic setting interface for the local area planning deleting

It is an opposite process of the local area planning adding compared with local area planning deleting. You need to delete the network planning first, and then the equipment. For example, you need to delete the local area and then the network planning if the local area has already been established.

If the color of the present area is orange, which means the local area is not established, the network planning can be deleted directly; if the color of the present area is blue, which means the local area has been established already, the local area need to be deleted in the object-tree or the command-tree. The exact steps goes as follows:

(1) Right-click the root directory of the network element equipment of the object tree, choose the 'network planning' in the 'dedicated function' and enter the local are a graphic setting interface, as shown in Figure 3-4-15.

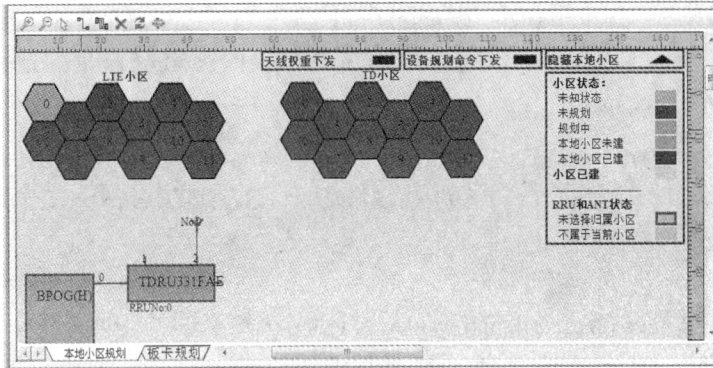

Figure 3-4-15　Local area planning interface for local area planning deleting

(2) Right-click the area 0 and choose to delete network planning, and then the local area begins to be deleted. When the area is successfully deleted, the successful shutting down information of the network element switch will be reported in the LMT message viewing window, as shown in Figure 3-4-16.

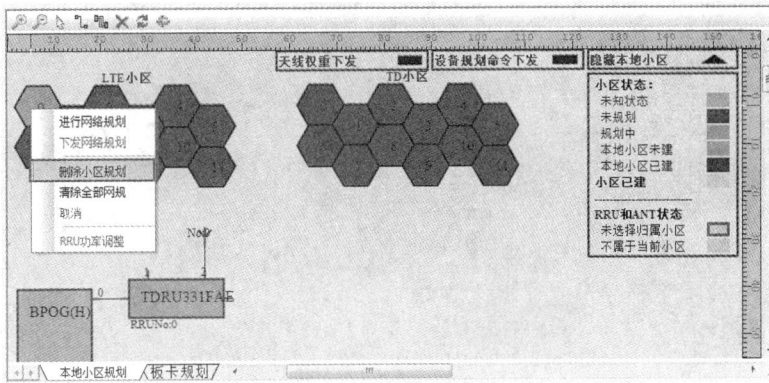

Figure 3-4-16　Area planning deleting interface

(3) Choose separately the antenna, RRU and BPOH which need to be deleted and press the 'Delete' key on your keyboard, and then all the chosen elements will be deleted; and also, you can choose all these elements, and then press the 'Delete' key to delete.

(4) Click the 'equipment planning order' to issue related orders, and make sure the deletion is working successfully. This is a step which is forgotten in lots of situations can cause failure and mistake for later area planning.

3. Graphic setting interface for modifying the local area planning

Modifying the area planning refers to making some adjustments to the local area planned before, for example, modifying the RRU type or antenna array type, etc. Due to the fact that the LMT can only support to add or delete the network planning operation, it is needed to delete all the network planning made before any modification and then to plan a new local area again.

(1) You need to delete all the network planning made before according to the instructions of deleting the local area planning in graphic setting interface, and you have to finish Step 4.

Specially, The network planning will definitely fail if you don't do what is instructed in Step 4.

(2) Plan a new area according to the instruction in Part 1: adding a local area planning in the graphic setting interface again.

Work Task 5: Local Area Parameter Setting

1. Setting the key parameters of an area

The init.cfg file has no area planning parameters until the V2.02.02.03 version is developed. If the base station has no cur.cfg file before and starts by using init.cfg file, the logic area cannot be established after the local area has been planned. You need manually add some key parameters for the area, such as the central frequency point, the frequency segment, the physical ID list of local area, the system bandwidth, etc.

(1) Choose the 'area parameters setting' in the setting management menu of the LMT menu bar, and open the area parameters setting interface, as shown in Figure 3-5-1.

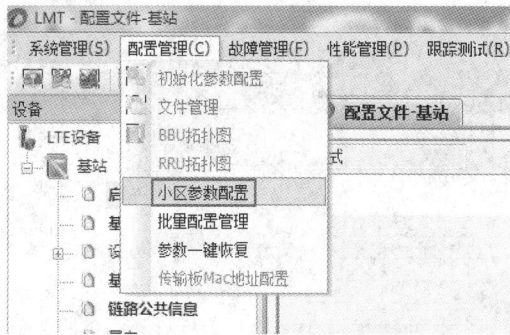

Figure 3-5-1   Area parameter setting interface

(2) Right click the area list to choose the 'area setting adding' order, as shown in Figure 3-5-2.

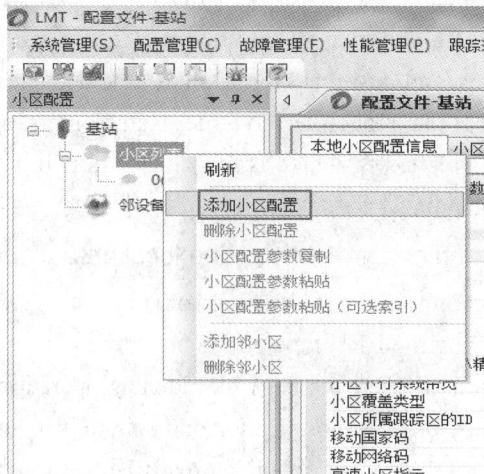

Figure 3-5-2   Area setting adding interface

(3) According to the area property interface, you need to finish the setting of related information for the area, such as, the local area ID, the area ID, the area friendly name, the area physical ID list, the E-UTRA working frequency, the area central frequency point, the area downlink system bandwidth, the area coverage type, the tracking ID belonged to area, etc. And then click to confirm, as shown in Figure 3-5-3.

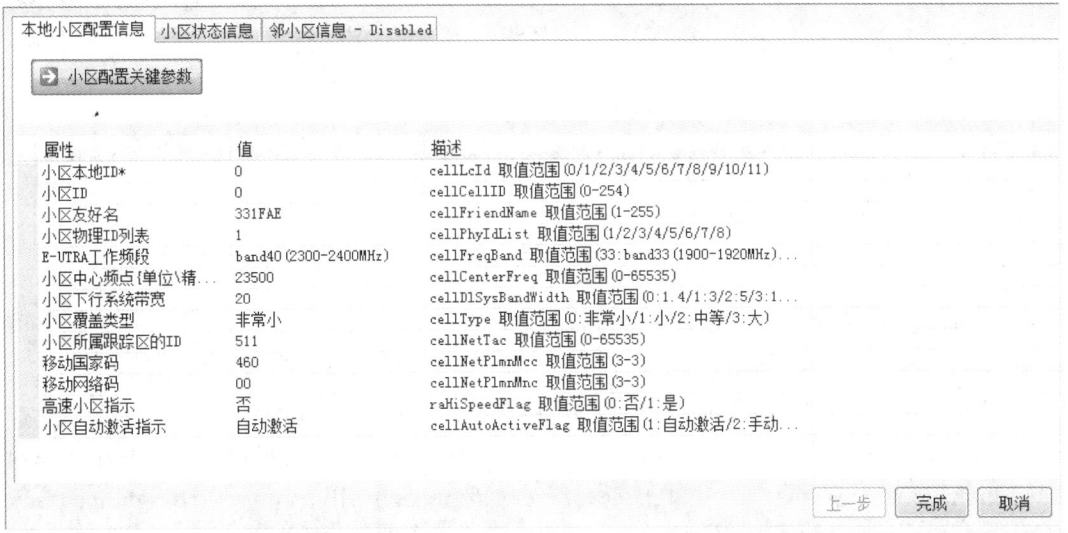

Figure 3-5-3   Local area parameter confirmation interface

You can activate the area after the setting of area parameters is finished.

Table 3-5-1   RRU work frequency

	Frequency	Carrier number	RRU
band40	2310 MHz	38750	RRU02
band40	2330 MHz	38950	RRU02 332E
band40	2350 MHz	39150	RRU02 332E
band38	2580 MHz	37850	318D
band38	2590 MHz	37950	318D
band38	2600 MHz	38050	318D
band38	2610 MHz	38150	318D
band38	2620 MHz	38250	318D
band32	2555 MHz	35800	318DL
band32	2560 MHz	35850	318DL
band32	2565 MHz	35900	318DL
band39	1910 MHz	37550	318FA

Table 3-5-2   work frequency description

Frequency Range	Frequency	Application Scenarios
A frequency range	2010–2025 MHz	Three indoor situation frequency points of TD-SCDMA + 6 outdoor coverage frequency points
E frequency range	2320–2370 MHz	Only used for indoor coverage
F frequency range	1880–1900 MHz	Outdoor coverage extended frequency range of TD-S CDMA system
F frequency range	1900–1920 MHz	Used for Personal Handphone System
D frequency range	2570–2620 MHz	LTE outdoor coverage
DL frequency range	2545–2575 MHz	Non-standard: meet the Japan SoftBank Cooperation requirement

Table 3-5-1 shows the area central frequency point, and Table 3-5-2 shows the correspondence relation among E-UTRA working frequencies. They are commonly used to modify the area setting parameters, and it's convenient to acquire the related parameters from the above Tables.

2. Area activation and deactivation operation

Find the 'TD-LTE service'→'TD-LTE area'→ 'area state information' in the object tree of LMT in the left of the interface, and the local area information will appear at the right interface. You can activate and deactivate the area when right-clicking to choose the 'area state information'→ 'activate (effective after 9 seconds)/deactivate area', as shown in Figure 3-5-4. You can get reference from Work Task 6 to solve the related problems when LMT print warning message pops up and the area cannot be activated during the related activation operation.

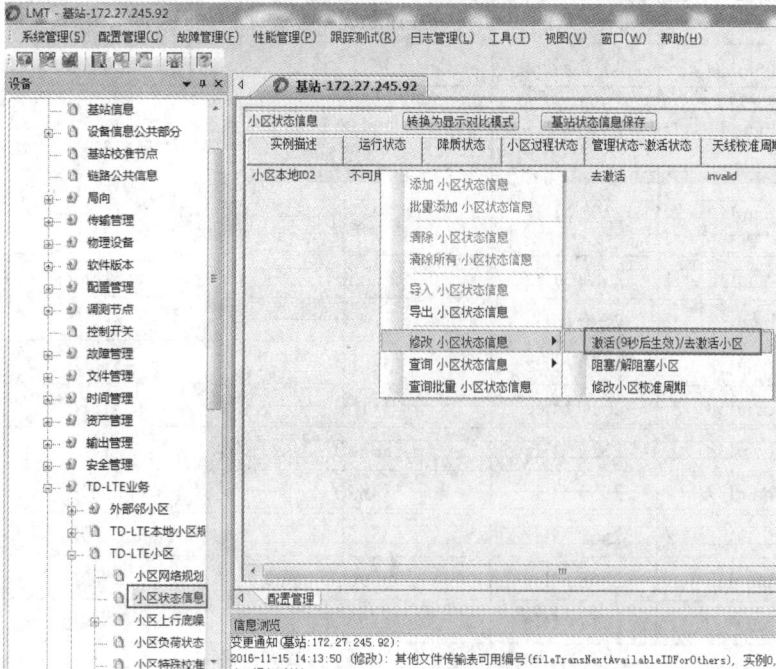

Figure 3-5-4　Area activation interface

Work Task 6: Common Problems and Solutions

You can check the starting process diagram to judge where the error is when the area cannot be activated, and it will be analyzed as follows:

(1) Login onto LMT and click the 'tool' in the menu, and open the 'start the process diagram' in the pull down menu, as shown in Figure 3-6-1.

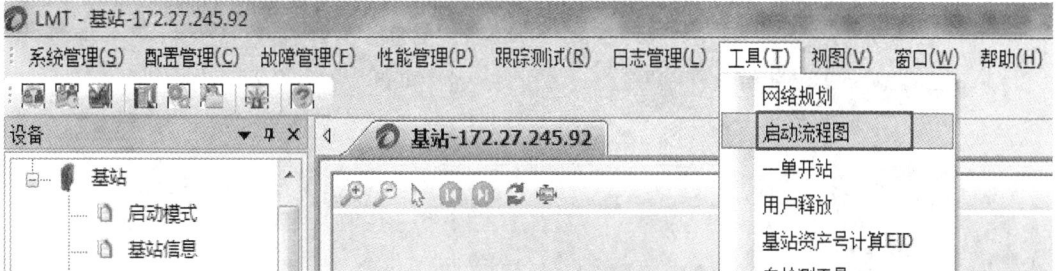

Figure 3-6-1   Tool menu

(2) The following picture shows the normal starting process. You need to pay attention to the TD-LTE (TDL) area state on the right, and the TD-SCDMA (TDS) state on the left could be ignored, the detailed starting process is as shown in Figure 3-6-2.

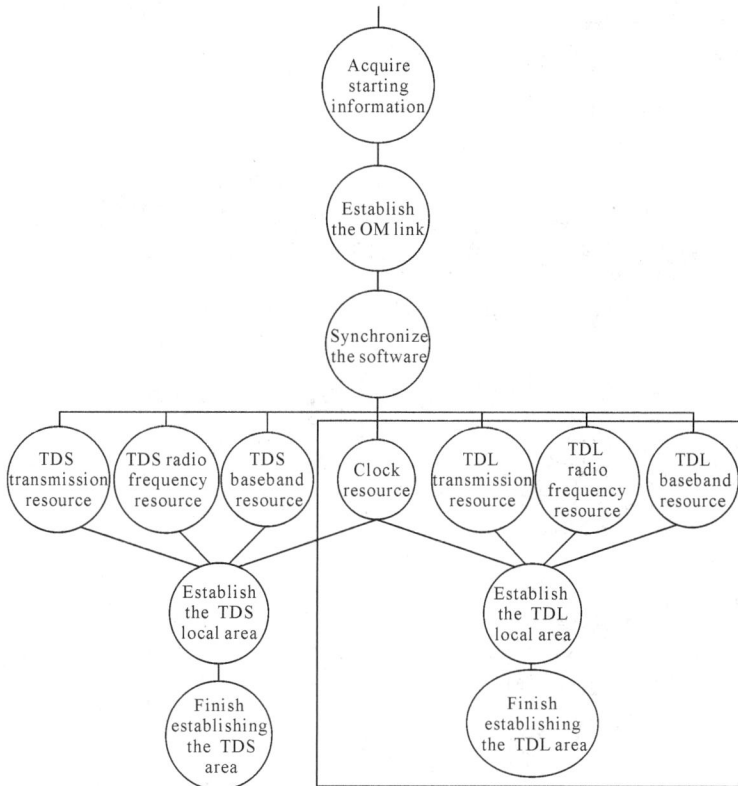

Figure 3-6-2   Starting process diagram

1. Clock resource failure

The failure is caused by GPS failure. The GPS is not available when the clock resource turns into red. In order to show the learners different equipment, the GPS is hanged indoors and cannot track the satellite. Thus, the clock resource cannot be accessed.

You can check and modify the clock mode to solve the problem:

(1) Check GPS satellite-lock status, and click LMT 'physical equipment' → 'clock information'. There are four clock checking statuses: clock source, present clock, GPS, far clock, all of which cannot be used when checked, as shown in Figures 3-6-3, 3-6-4 and 3-6-5.

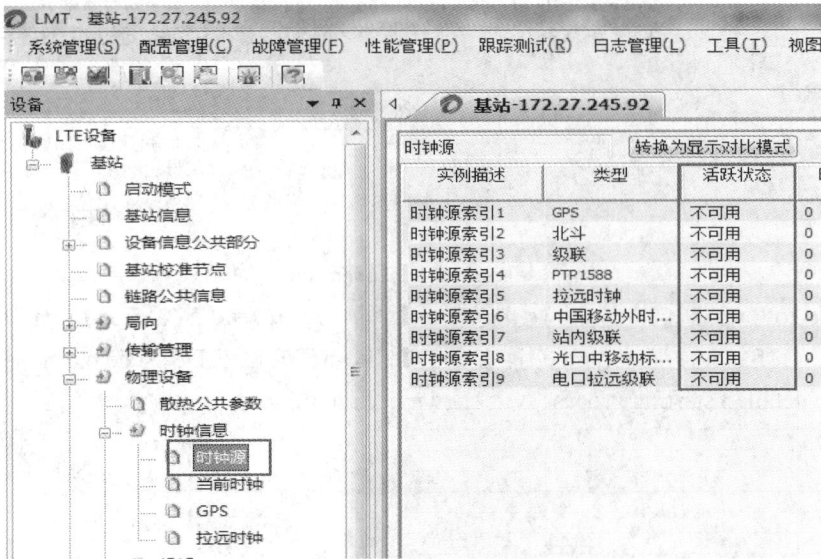

Figure 3-6-3　Physical equipment interface

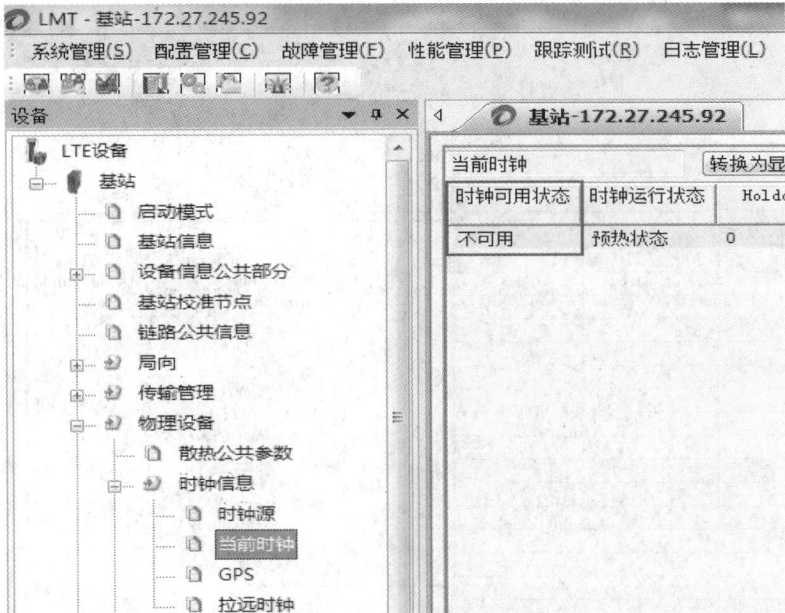

Figure 3-6-4　Current clock interface

LMT - 基站-172.27.245.92

系统管理(S) 配置管理(C) 故障管理(F) 性能管理(P) 跟踪测试(R) 日志管理(L) 工具(T) 视图(V) 窗口(W) 帮助(H)

设备 ▼ 🗗 × ◁ ● 基站-172.27.245.92

LTE设备
基站
　启动模式
　基站信息
　设备信息公共部分
　基站校准节点
　链路公共信息
　局向
　传输管理
　物理设备
　　散热公共参数
　　时钟信息
　　　时钟源
　　　当前时钟
　　　GPS
　　　拉远时钟

GPS　　转换为显示对比模式　　基站状态信息保存

实例描述	GPS设备类型	经度	纬度	海拔	故障原因	锁星数
机架0机框0插槽1GPS设备编号0	GPS	121.58147	29.85942	49	未跟踪到卫星	0

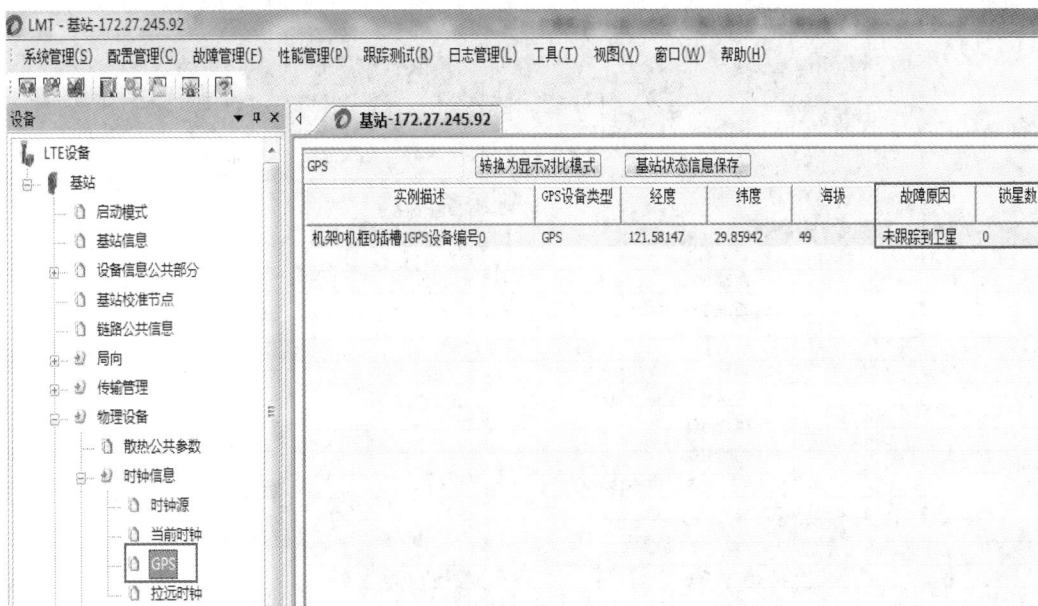

Figure 3-6-5　GPS clock interface

(2) The problem can be solved as follows: click the 'starting mode', and modify the 'starting mode of system clock source' into 'no-clock source mode', as shown in Figure 3-6-6. (Attention: if the base station reports the command response is overtime after it finishes modifying the mode, it is because it doesn't modify the clock source mode in time after the base station starts, and the clock source enters the 'holdover' state. You can restart the base station and modify the starting mode of the clock source as quickly as possible, and the problem will be solved.)

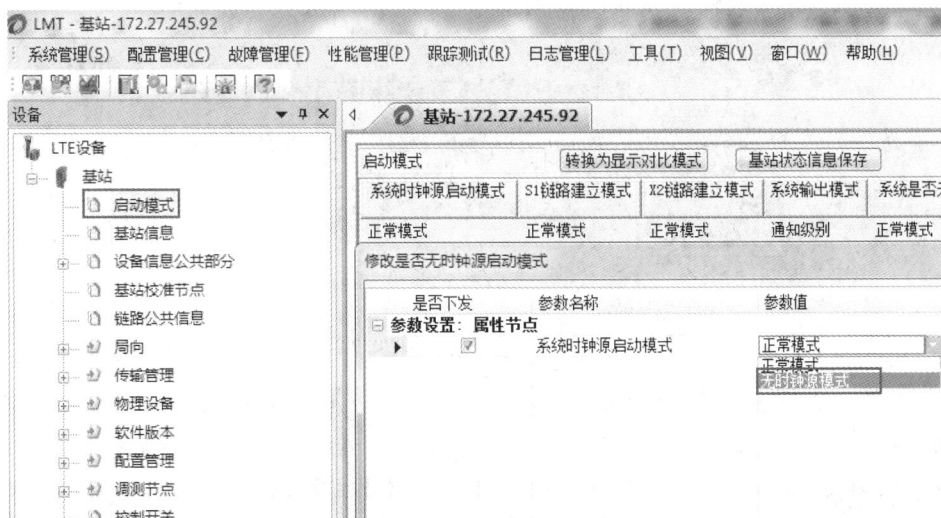

LMT - 基站-172.27.245.92

系统管理(S) 配置管理(C) 故障管理(F) 性能管理(P) 跟踪测试(R) 日志管理(L) 工具(T) 视图(V) 窗口(W) 帮助(H)

设备 ▼ 🗗 × ◁ ● 基站-172.27.245.92

LTE设备
基站
　启动模式
　基站信息
　设备信息公共部分
　基站校准节点
　链路公共信息
　局向
　传输管理
　物理设备
　软件版本
　配置管理
　调测节点
　控制开关

启动模式　　转换为显示对比模式　　基站状态信息保存

系统时钟源启动模式　S1链路建立模式　X2链路建立模式　系统输出模式　系统是否于

正常模式　　　正常模式　　　正常模式　　　通知级别　　　正常模式

修改是否无时钟源启动模式

是否下发　　参数名称　　　　　参数值
参数设置：属性节点
　　☑　　系统时钟源启动模式　　正常模式
　　　　　　　　　　　　　　　　正常模式
　　　　　　　　　　　　　　　　无时钟源模式

Figure 3-6-6　Starting mode interface

You can use the clock source after modification, and the clock source in the starting process diagram turns to green. The state of current clock will be 'available' and in 'locked' state, when you check it, as shown in Figure 3-6-7.

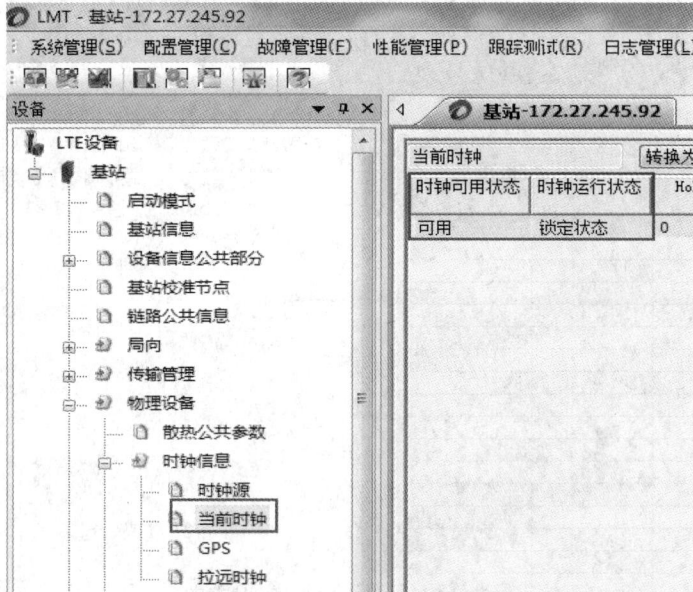

Figure 3-6-7   Current clock interface

2. TDL transmission resource failure

The S1 link failure can cause TDL transmission resource failure. You can check the link public information and find S1 link in failure state, as shown in Figure 3-6-8.

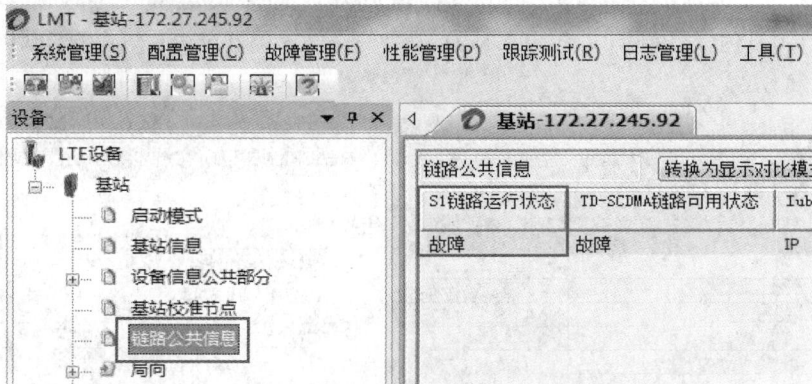

Figure 3-6-8   Link public information interface

This problem can be solved by the steps as follows.

(1) Check the physical link state. You need to check whether the lights on the back board MCPB and the core network box GECB are right or wrong. If the lights do not work, you probably need to wait about 30 minutes due to the possibility that the core network is on starting process. You need to use another net cable if the lights do not work all the time.

(2) You can set the IP to '172.1.1.200' in the local PC when the net light is on. You need to open the PC command prompt and input: ping 172.1.1.100, ping 172.1.1.2, and then observe

whether the Ping connection is successful or not, as shown in Figure 3-6-9. Go to Step 3 if Ping connection is successful, and ask for help from Datang Mobile technicians if the Ping connection fails.

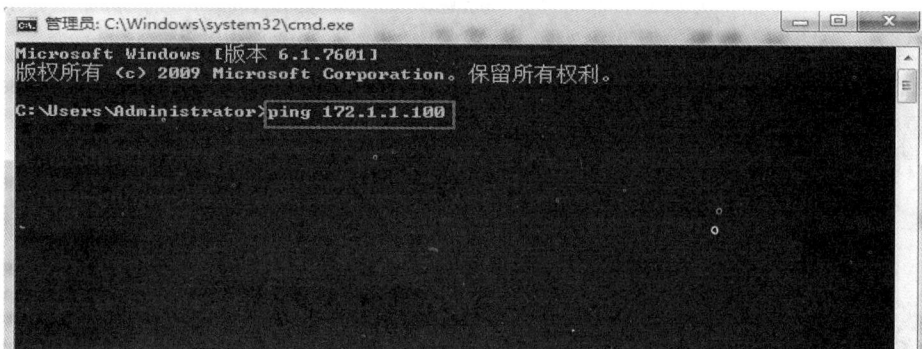

Figure 3-6-9   Link testing interface

(3) Check the transmission setting in the base station, which can be divided into the following parts:

① Check network element identification (Logic ID), click 'Base station'and observe whether the network element identification is 0, as shown in Figure 3-6-10.

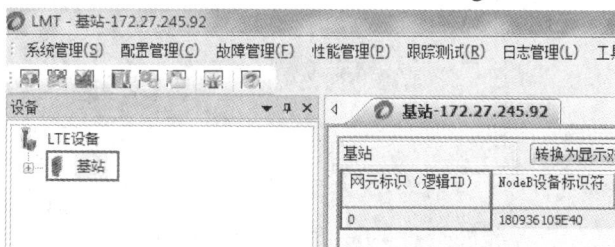

Figure 3-6-10   Network element identifying interface

② Check the area tracking domain ID, the mobile country code, and the mobile network code. You need to click 'TD-LTE services'→ 'TD-LTE area'→ 'area network planning', and observe whether the three values are 511, 123 and 11 respectively, as shown in Figure 3-6-11.

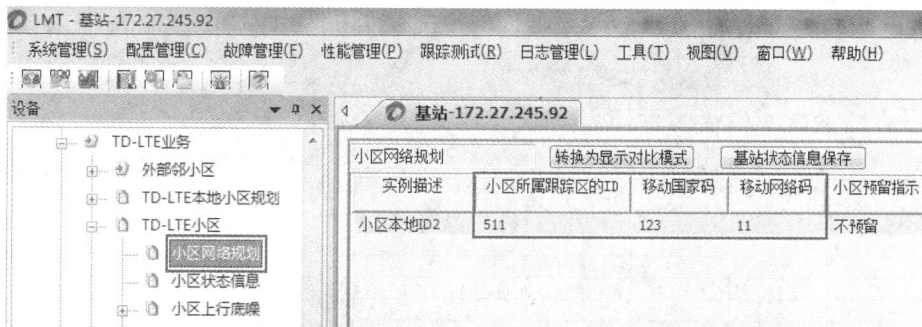

Figure 3-6-11   Values Checking at the area network planning interface

③ Check the base station IP address, click 'transmission management'→ 'IP address', and check whether the address is 172.1.1.99 and the numbers of the physical interfaces are the same as in Figure 3-6-12.

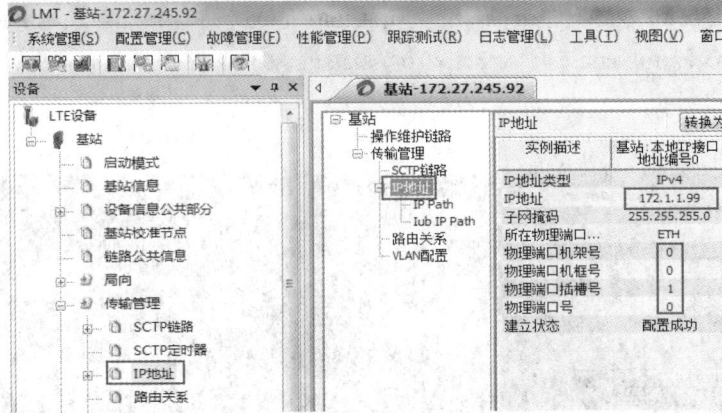

Figure 3-6-12　IP address interface

④ Check the SCTP link, click 'transmission management'→'SCTP link', and check whether the counter interface IP address 1 is 172.1.1.2 and the SCTP link establishment status is shown as 'successfully connected with the counter interface', as shown in Figure 3-6-13.

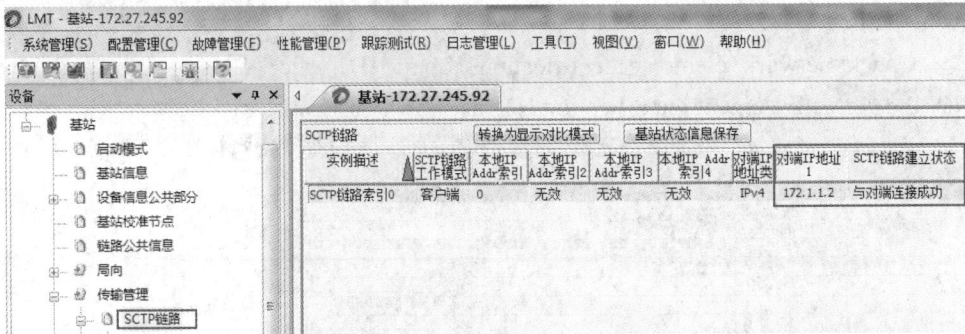

Figure 3-6-13　SCTP link interface

⑤ All the events related in steps 1-4 are normal, but the SCTP link establishment status shows the drivers and high-level settings are successful. You can try to issue transmission cut-off and connect the SCTP link manually. Click 'transmission management'→ 'transmission cut-off', right-click in the blank area, choose 'modify transmission cut-off' order, click 'make TD-LTE cut-off effective', and then wait for the successful connection, as shown in Figure 3-6-14.

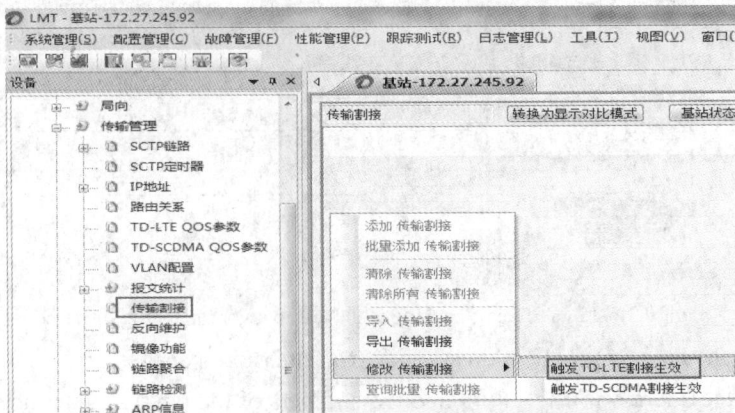

Figure 3-6-14　Transmission cut-off interface

⑥ Check the link public information through LMT after the transmission link modification is finished, and find that S1 link operation status is normal, as shown in Figure 3-6-15.

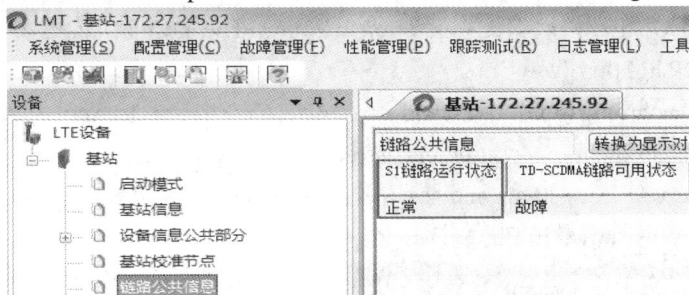

Figure 3-6-15   Link public information interface

3. TDL radio frequency resource failure

This failure is caused by unsuccessful connection of RRU. When RRU cannot be connected, you cannot get the RRU information in the LMT 'physical equipment'→ 'radio frequency unit topology', as shown in Figure 3-6-16.

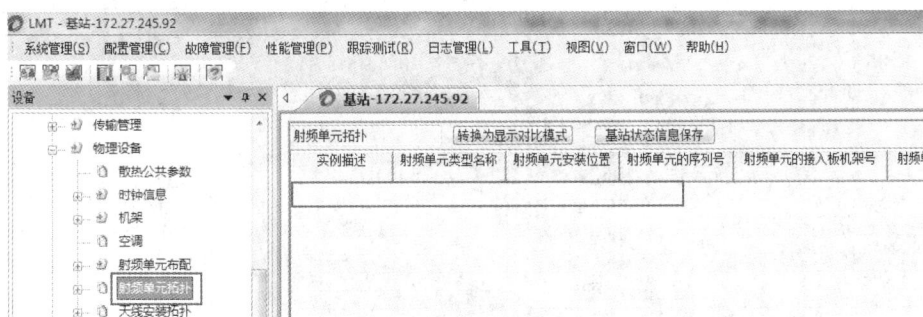

Figure 3-6-16   Radio frequency unit topology interface

You can make reason checking for RRU connection failure as follows.

1) No power for RRU

Check whether RRU has power. You need to check whether the switch of lightening protective box is turned on, and open the RRU operation window if the previous check is confirmed. You need to check the working condition of RRU power light PWR, and RRU has power if PWR is always on, as shown in Figure 3-6-17.

Figure 3-6-17   RRU interface circuit

2) Damage to the optical module and the optical fiber

You can judge whether the working statue of optical module by checking the IR light on BPOH board and OP light in RRU operation window. The normal status is that the IR light flickers and the OP light is always on.

Check the synchronization status and optical receiving and sending status through LMT, click 'physical equipment' → 'equipment structure' → 'equipment box' → 'board' → 'optical module', and observe FPGA status and service status, which should be 'synchronization' and 'in service' normally, as shown in Figure 3-6-18.

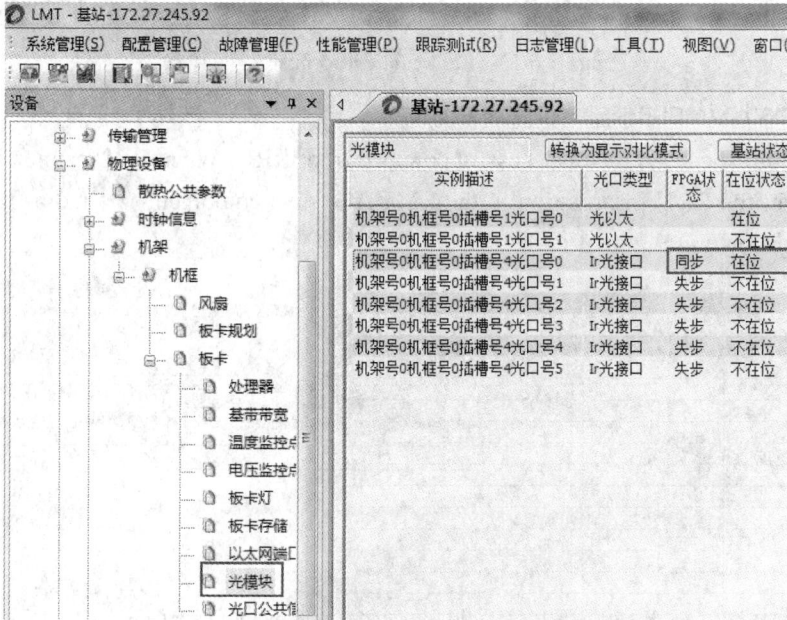

Figure 3-6-18   Optical module interface

If it is sure that the damage to the optical module and the optical fiber has caused unsuccessful connection of RRU, you should replace the related optical module or optical fiber. Be careful when replacing the optical fiber, and please do not bend the optical fiber too much to avoid damage.

3) Wrong planning and setting of network elements

During network element planning and setting, the failure is caused by the mismatching between RRU planning and its real situation. You will find a warning from the mismatching between RRU and its real situation when you check LMT active warnings. The solution is to plan the related network elements according to its real situation again, and you can get reference from Work Task 4.

4) Unsuccessful connection after broadcasting connection request sent by RRU

You can find RRU's repeated requests for connection by message printed LMT when this failure appears, and it is connected again until the broadcasting connection completes. The solution is to restart RRU for a successful reconnection in the end, as shown in Figure 3-6-19. (Attention: upload all RRU logs when this situation happens and contact Datang Mobile technicians for related analysis.)

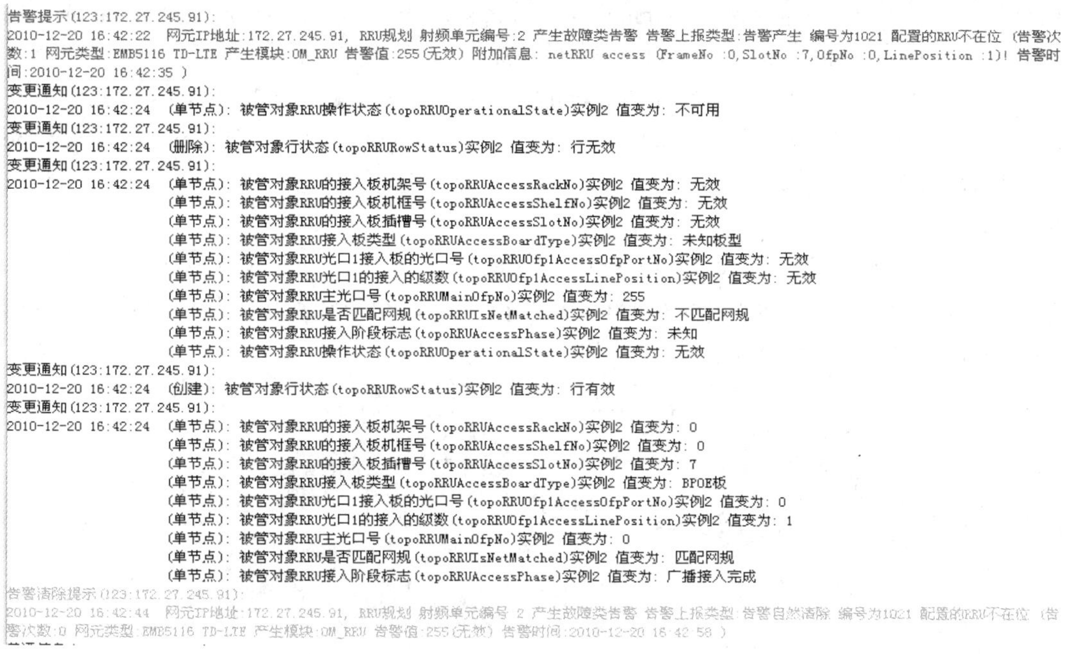

```
告警提示(123:172.27.245.91):
2010-12-20 16:42:22  网元IP地址:172.27.245.91, RRU规划 射频单元编号:2 产生故障类告警 告警上报类型:告警产生 编号为1021 配置的RRU不在位 (告警次
数:1 网元类型:EMB5116 TD-LTE 产生模块:OM_RRU 告警值:255(无效) 附加信息: netRRU access (FrameNo :0,SlotNo :7,OfpNo :0,LinePosition :1)! 告警时
间:2010-12-20 16:42:35 )
变更通知(123:172.27.245.91):
2010-12-20 16:42:24  (单节点): 被管对象RRU操作状态(topoRRUOperationalState)实例2 值变为: 不可用
变更通知(123:172.27.245.91):
2010-12-20 16:42:24  (删除): 被管对象行状态(topoRRURowStatus)实例2 值变为: 行无效
变更通知(123:172.27.245.91):
2010-12-20 16:42:24  (单节点): 被管对象RRU的接入板机架号(topoRRUAccessRackNo)实例2 值变为: 无效
                     (单节点): 被管对象RRU的接入板机框号(topoRRUAccessShelfNo)实例2 值变为: 无效
                     (单节点): 被管对象RRU的接入板插槽号(topoRRUAccessSlotNo)实例2 值变为: 无效
                     (单节点): 被管对象RRU接入板类型(topoRRUAccessBoardType)实例2 值变为: 未知板型
                     (单节点): 被管对象RRU光口1接入板的光口号(topoRRUOfp1AccessOfpPortNo)实例2 值变为: 无效
                     (单节点): 被管对象RRU光口1的接入的级数(topoRRUOfp1AccessLinePosition)实例2 值变为: 无效
                     (单节点): 被管对象RRU主光口号(topoRRUMainOfpNo)实例2 值变为: 255
                     (单节点): 被管对象RRU是否匹配网规(topoRRUIsNetMatched)实例2 值变为: 不匹配网规
                     (单节点): 被管对象RRU接入阶段标志(topoRRUAccessPhase)实例2 值变为: 未知
                     (单节点): 被管对象RRU操作状态(topoRRUOperationalState)实例2 值变为: 无效
变更通知(123:172.27.245.91):
2010-12-20 16:42:24  (创建): 被管对象行状态(topoRRURowStatus)实例2 值变为: 行有效
变更通知(123:172.27.245.91):
2010-12-20 16:42:24  (单节点): 被管对象RRU的接入板机架号(topoRRUAccessRackNo)实例2 值变为: 0
                     (单节点): 被管对象RRU的接入板机框号(topoRRUAccessShelfNo)实例2 值变为: 0
                     (单节点): 被管对象RRU的接入板插槽号(topoRRUAccessSlotNo)实例2 值变为: 7
                     (单节点): 被管对象RRU接入板类型(topoRRUAccessBoardType)实例2 值变为: BPOE板
                     (单节点): 被管对象RRU光口1接入板的光口号(topoRRUOfp1AccessOfpPortNo)实例2 值变为: 0
                     (单节点): 被管对象RRU光口1的接入的级数(topoRRUOfp1AccessLinePosition)实例2 值变为: 1
                     (单节点): 被管对象RRU主光口号(topoRRUMainOfpNo)实例2 值变为: 0
                     (单节点): 被管对象RRU是否匹配网规(topoRRUIsNetMatched)实例2 值变为: 匹配网规
                     (单节点): 被管对象RRU接入阶段标志(topoRRUAccessPhase)实例2 值变为: 广播接入完成
告警清除提示(123:172.27.245.91):
2010-12-20 16:42:44  网元IP地址:172.27.245.91, RRU规划 射频单元编号 2 产生故障类告警 告警上报类型: 告警自然清除 编号为1021 配置的RRU不在位 (告
警次数:0 网元类型:EMB5116 TD-LTE 产生模块:OM_RRU 告警值:255(无效) 告警时间:2010-12-20 16:42:58 )
```

Figure 3-6-19   Warning indication interface

Here are the functions of the indication lights in the RRU operation window:

PWR: power indication light

VSWR: standing-wave ratio indication light

CLK: clock indication light

OP1: optical fiber connection interface 1 indication light

OP2: optical fiber connection interface 2 indication light

The three lights of PWR,VSWR and CLK are always on when RRU works normally, and the indication lights for various optical fiber connection interfaces are always on as well according to the working conditions.

4. TDL baseband resource failure

This failure happens when the BPOH board fails or is on starting, and you need to wait about 10 minutes after the base station starts, under such condition could the BPOH board have a normal starting.

If it is sure that the board cannot start, you need to observe the ALM light status of the BPOH board. The normal state of the light is off, and it will be on and with red color when failures develop. A flickering red light means the failure can be solved, but the hardware may have some problems when the red light is always on, which means the problem cannot be solved. You need to ask Datang Mobile technicians for help in this situation.

Solutions to the solvable failures：

(1) Wrong network element planning and settings can lead to BPOH board failure, and

there will be a warning for the mismatching between the board type and its real situation. The solution is to adjust the network planning and settings again according to the real situation, and related operation method can be found in Work Task 3.

(2) The file loss of the BPOH board can lead to BPOH failure, which doesn't influence area activation. There will be no radio frequency signal and the terminal cannot search the LTE wireless signal. The solution is to unpack for updating, and you can get reference from Work Task 4.

You can check the board status through LMT after the problem is solved, and click 'physical equipment'→'equipment structure'→'equipment box'→'board'. The normal status of the BPOH board is shown in Figure 3-6-20.

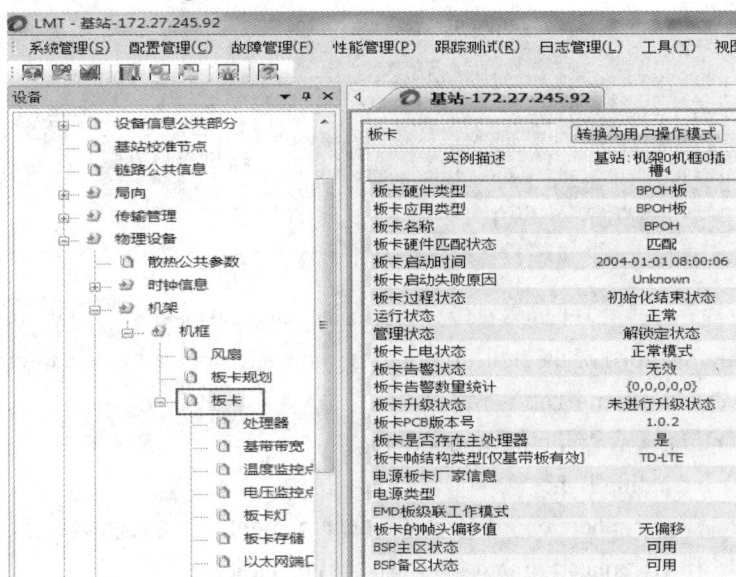

Figure 3-6-20   Board message interface

5. Local area activation failure

A successful establishment of the local area means that the previously mentioned clock resource, the transmission resource, the radio frequency resource and the baseband resource are available. Local area activation failure is mainly caused by the mismatching between TD-LTE area setting and its planning, and LMT prints the warning message during the activation process. We can judge which part of the parameter setting is wrong according to the warning message.

The following are some common and typical examples, and you can solve the problem according to the file indication when the same warning appears. If you cannot find the related reason for activation failure in the document, you should contact Datang Mobile technicians for help.

1) Central frequency point RRU unable to support frequency segment setting of TD-LTE area

The following warning message will be displayed if the situation happens: "Cell(2)'s

CenterFreq does not match with the RRU's capability", as shown in Figure 3-6-21.

Figure 3-6-21   CenterFreq Warning interface

Solution: check the frequency segment supported by RRU, and confirm that the product instruction from the RRU operation window: all the frequency segments are supported by RRU.

After the confirmation, you need to modify the TD-LTE area frequency segment and central frequency point in LMT: click 'TD-LTE service'→ 'TD-LTE area', and right-click to choose 'modify TD-LTE area'→'modify area setting parameter', as shown in Figure 3-6-22.

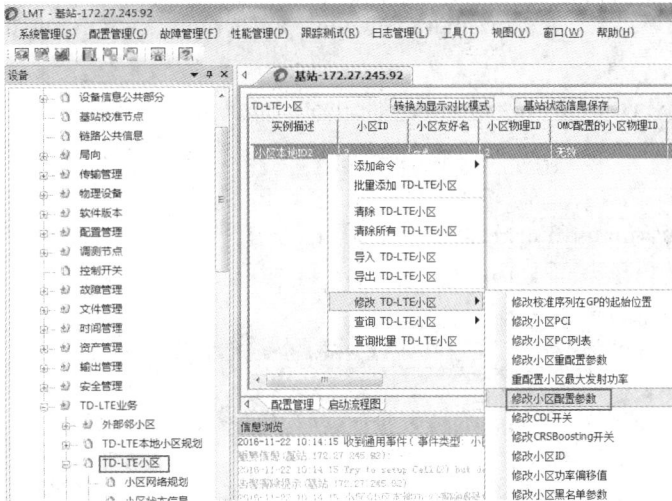

Figure 3-6-22   Area parameter modifying interface

Modify the E-UTRA working frequency range and the area central frequency point so that the range will be supported by RRU, and then activate the area, as shown in Figure 3-6-23.

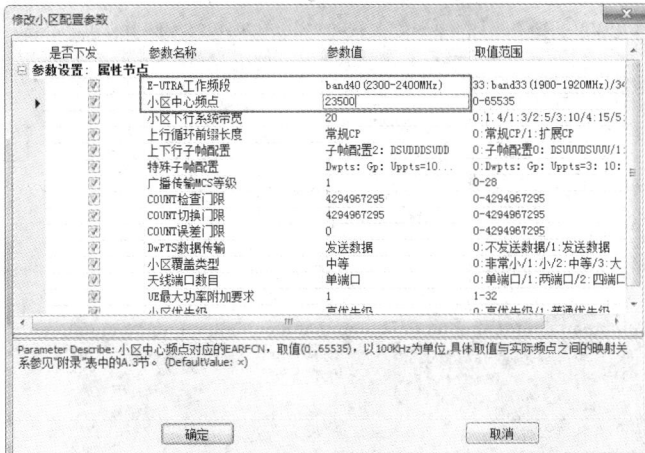

Figure 3-6-23   Area parameter setting interface

2) The maximum transmitting power of the local area is larger than the maximum downlink power of the local area.

The following warning message will be displayed if the situation happens: "cell (2)'s MaxPwr is larger than localcell!", as shown in Figure 3-6-24.

Figure 3-6-24   Warning message interface

Solution: check the maximum downlink power and the minimum downlink power of the local area in LMT, and confirm to modify the maximum transmitting power of the TD-LTE area. Its value should be between the maximum downlink power and the minimum downlink power of the local area.

Check the maximum downlink power and the minimum downlink power of the local area: click 'TD-LTE service'→'TD-LTE local area planning'→'TD-LTE local area', as shown in Figure 3-6-25.

Modify the maximum transmitting power of the TD-LTE area: click'TD-LTE service'→ 'TD-LTE area', and right-click to 'modify TD-LTE area'→ 'reset the maximum transmitting power of area', as shown in Figure 3-6-26.

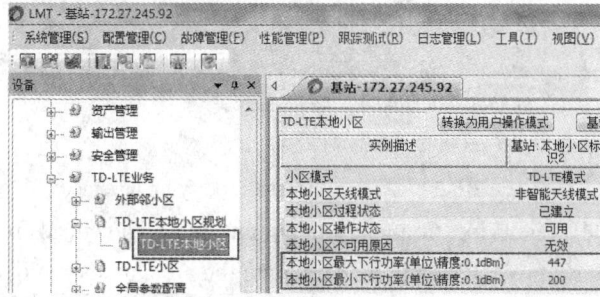

Figure 3-6-25   Local area transimitting power checking interface

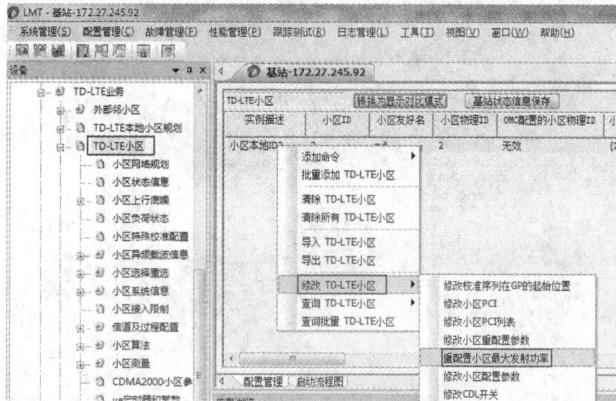

Figure 3-6-26   Area parameter modifying interface

3) Antenna mode setting does not match antport number (at antenna interface)

Print the additional message of warning: "Cell(2)'s Transmission mode does not match with antport number!", as shown in Figure 3-6-27.

Figure 3-6-27　Transmission mode warning mode interface

This failure is caused by using an indoor non-intelligent antenna, which only supports TM1 antenna mode. This kind of failure will appear due on the condition that the default setting is TM3 antenna mode when we start the base station for the first time.

Solution: modify the antenna mode in LMT, click 'TD-LTE service'→ 'TD-LTE area'→ 'signal channel and process setting'→ 'antenna parameter', and right-click to 'modify antenna parameter'→'modify downlink transmission mode'. What you need to do is to modify the value into TM1, then the problem will be solved, as shown in Figure 3-6-28.

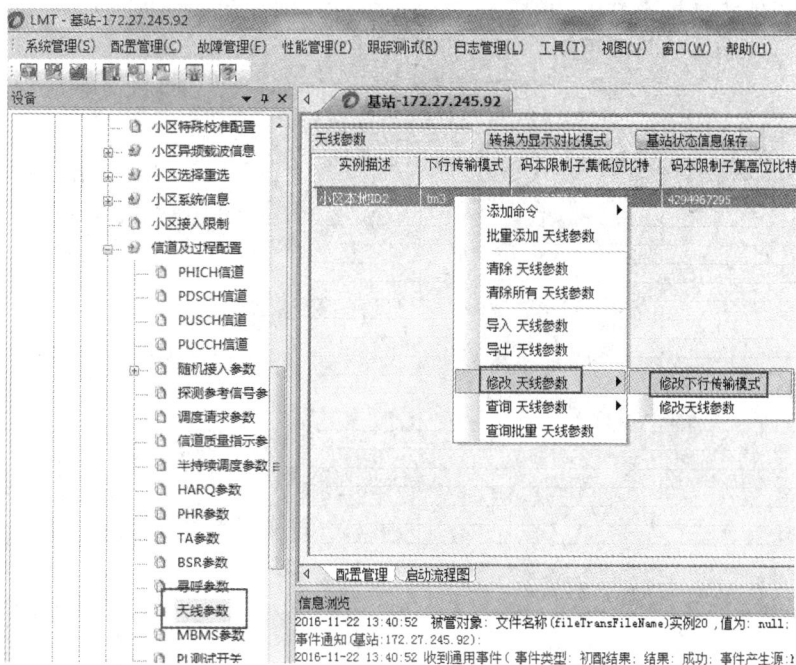

Figure 3-6-28　Antenna parameter modification interface

References

[1]   陈宇恒，肖竹，王洪. LTE 协议栈与信令分析[M]. 北京：人民邮电出版社，2013.

[2]   王映民. TD-LTE 技术原理与系统设计[M]. 北京：人民邮电出版社，2010.

[3]   杨丰瑞，文凯，吴翠先. LTE/LTE-Advanced 系统架构和关键技术[M]. 北京：人民邮电出版社，2015.

[4]   中国通信建设集团设计院有限公司. LTE 组网和工程实践[M]. 北京：人民邮电出版社，2014.

[5]   大唐移动通信设备有限公司. TD-LTE 虚拟实验室 V-Lab 软件指导手册.

[6]   尹圣君，钱尚达，李永代，等. LTE 及 LTE-Advanced 无线协议[M]. 张鸿涛，   等，译. 北京：机械工业出版社，2015.

[7]   李正茂，王晓云. TD-LTE 技术与标准[M]. 北京：人民邮电出版社，2013.

[8]   高峰，高泽华，丰雷，等. TD-LTE 技术标准与实践[M]. 北京：人民邮电出版社，2011.

[9]   元泉.   LTE 轻松进阶[M]. 北京：电子工业出版社，2012.

[10]   谢显忠. 基于 TDD 的第四代移动通信技术[M]. 北京：电子工业出版社，2005.